Curiosity Unleashed

QUESTIONS AND ANSWERS ABOUT THE NATURAL WORLD

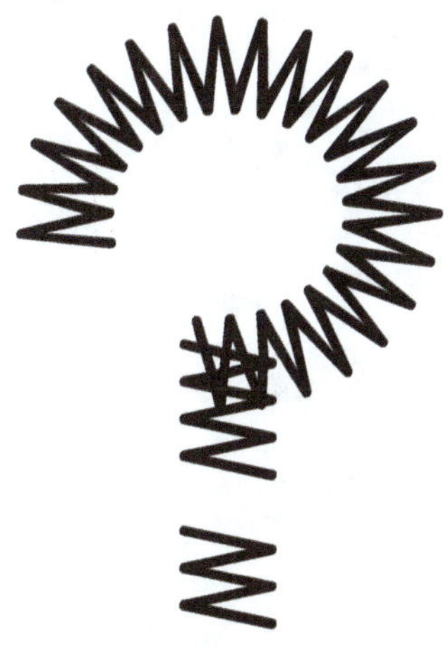

DEREK EAMUS

ISBN: 978-1-7636785-0-7 (print)
       978-1-7636785-1-4 (ePub)

Published by: Pole-House Books

Interior design by Booknook.biz

PREFACE

Why did I write this book? A good question.

I have two answers. First, it is because asking questions about what I see around me has always been, and continues to be, central to my life. The first question I remember asking my Dad, almost 60 years ago, was 'Why is the sky blue?' (Chapter 10). He didn't know the answer, but we found it in the new set of Encyclopedias that my parents had recently purchased. And that was it. Asking questions, and seeking answers, was the most exciting thing to do – ever. For 40 years, I pursued that excitement in the most fantastic job – being a research scientist, lecturer, mentor and eventually Professor. And writing this book has let me do the best parts of this: ask questions, pursue answers, and hopefully, possibly, engage others in those questions and answers. I get a real buzz from revealing links among topics you never realised were linked and uncovering that **WOW!** moment.

The Wow moments, for me, come when I ask a simple question – and then discover that the initial answer leads into some unexpected places, into some very unexpected insights. This point *explains the structure of each chapter*. At the start of each chapter, I ask a simple question – for example, why does ice float (Chapter 3)? After answering that simple question, I discovered that if ice did not float, life on Earth would be very, very, different. I then discovered that Earth has been almost entirely covered in snow and ice several times (Snowball Earth; Chapter 3). Similarly, in answering the question: 'Why doesn't the Moon fall to Earth?' I discovered the cause of the extinction of the dinosaurs 65 million years ago and I also discovered that jellyfish exert a very large force on oceans when they swim *en masse*.

The second, and minor reason for writing this book was to provide a counter-narrative to some of the nonsense I read and hear spoken about Science in general, and climate change in particular. A sort of 'myth busting'. I have received a lot of research funding to examine how plants respond to climate change and have had a long-term and in-depth interest in the Science of climate change. And frankly, what is said in the media about climate change is frequently unadulterated twaddle. So, I spend a reasonable amount of time in chapters 11 and 12 looking at the causes of climate change, the impacts of climate change on the frequency and intensity of catastrophic events – droughts, wildfires, tropical cyclones, and frequency and intensity of El Niño and La Niña (Chapter 11), and a discussion of the questions: 'What is Science?' and 'Is the Science of climate change settled?' (Chapter 12).

So, I hope there is a **WOW!**, or even several, for each reader (both of you) as you randomly dip in and out of each chapter.

Enjoy!

Derek Eamus
Emeritus Professor

Table of Contents and Abstracts

Preface iii

Chapter 1 **What are jet trails and how are they made? What is the link between the 9/11 attack on the Twin Towers, jet trails, and land surface temperatures?** 3

On the morning of the 9th of September 2001, four passenger airlines were hijacked by Islamic terrorists of the al-Qaeda group. One jet was deliberately flown into the North tower of the World Trade Centre (WTC) in New York, another was flown into the South tower of the WTC, and a third jet was flown into the Pentagon in Virginia, USA. A fourth jet crashed into a field in Pennsylvania after a determined defence of the plane by passengers on board.

In response to the attack, all civilian flights into, out of, and across the USA were cancelled for three days. This was an unprecedented response to an unprecedented attack. But it provided a unique opportunity to test an important hypothesis: that contrails can alter the weather.

After looking at how contrails are made, and the importance of contrails to the surface energy balance of a landscape, this chapter then shows how the eruption of Mount Pinatubo in 1991 resulted in a massive increase in the rate of uptake of CO_2 into trees – so much so that atmospheric CO_2 levels stopped increasing for a year or two. The specific example of contrails and their impact on temperature and weather is merely one example of global dimming, whereby the amount of solar radiation arriving at the Earth's surface has declined for several decades. However global dimming has now been replaced by global brightening. How do global dimming and global brightening affect carbon uptake and water use by vegetation?

A common assumption made in the media and some scientific publications is that as the world warms up, the rate of evaporation from lakes, wet soil surfaces, and vegetation must increase, in the same way that a hot pan of water on a cook top evaporates more rapidly than a cool pan of water sitting on a bench top. However, the Pan Paradox states that, to the contrary of this widely held assumption, evaporation from open surfaces of water have been declining for the past 50 years, despite temperatures rising over that period. This paradox is discussed and put into the context of regional water budgets.

An absolute cornucopia of subjects to be addressed in this chapter.

Chapter 2   Why doesn't the Moon fall to Earth? An exploration of gravity, tides, jellyfish, and the Moon (and why we never see both sides of the Moon, despite the Moon spinning on its axis). 25

We all know the Moon makes the tides on Earth, right? But the Sun is 27 million times heavier than the Moon and only 390 times further away – so does the Sun have any influence on our tides? And why do we have two high and two low tides per day but only one moon? How do jellyfish come into all this? And what can satellites measuring small differences in the gravity field around Earth tell us about water on Earth? This chapter tackles these issues and reveals how new satellites are being used to look at changes in the Earth's gravity and water supplies. This chapter also addresses the questions: 'Why doesn't the Moon fall to Earth – and what would happen if it did?' Oh yes, finally, as an aside, I answer the question: If the Moon is spinning, why do only ever see the same side whenever we look up at the Moon?

Chapter 3   Why does ice float and how has this affected the evolution of life on Earth? 47

Ice (frozen water) floats. We all know that. The North Pole (the Arctic) is an island of ice floating on seawater. Ice in my vodka floats. However, most liquids, when frozen, become more dense and sink when dropped into the liquid form. But not water. Why not? Perhaps more importantly, what would happen if ice didn't float? Turns out that there would be very profound impacts on life on Earth because Earth has indeed frozen almost entirely three times in the past 2.4 billion years (when Earth was known to be a slushy ice ball or a snowball). Why did Earth freeze over three times in the past 2.4 billion years, And are we really currently living through an ice age? These are some of the key questions to be examined in this chapter.

Chapter 4   What is the freezing point of pure water, and why don't conifers freeze to death in winter? Can we really cryonically freeze humans for later revival if several animal species can be frozen and thawed back to life? 67

We all know water freezes at 0 °C, right? This was certainly assumed in Chapter 3. However, in that discussion I was talking about our common experience of the freshwater we find in our taps, in rivers, lakes and non-saline groundwater. When we discuss *absolutely* pure water, however, perhaps 0 °C is not the correct answer. Everyday measurements of everyday water demonstrate that a 0 °C freezing temperature for water suffers from the presence of two important artefacts, which I discuss later in this chapter. So, what is the freezing point of pure water when these two artefacts are removed?

Water does undoubtedly freeze, and many species of plants and animals must survive extremely cold winters without dying, despite containing a lot of water. How do they stay alive? What is their secret? Indeed, how do some animals survive being frozen solid and yet 'come-back-to-life' after thawing? Can this teach us anything about cryonics, the process of freezing and preserving tissues (especially human ones) and bringing them 'back-to-life'?

What an eclectic mix of themes tackled in this chapter.

Chapter 5   What is groundwater? How did groundwater influence early human evolution? What is over-extraction of groundwater doing to Jakarta and northern China? 85

Groundwater is the vast, mostly hidden, store of water lying underground. More than 2 billion people globally rely on groundwater for drinking and within urban areas, about 50 % of human consumptive water use comes from groundwater stores. Of the approximately 750 km^3 (750,000 billion litres) of water extracted from groundwater resources each year, about 70 % is used for agricultural purposes. But how far back in history can we trace groundwater use? Did groundwater access affect early human evolution in Africa? Has groundwater scarcity caused civil unrest and led to armed conflict? How has access to groundwater become weaponised? And, where do we find groundwater, how is it different from just plain old soil water, and does groundwater get replenished, and if so, how? Does it matter to the ecology of a region if groundwater is extracted too quickly? Does anything live and breed in aquifers? How can two satellites help us manage groundwater and why is China moving nearly 30 km^3 (30,000 billion litres) of water *annually* from southern China to northern China? Finally, what is overextraction of groundwater doing to Jakarta and northern China? These are the key questions I address in this chapter.

Chapter 6   What makes the Earth's magnetic field? Why do we care about our magnetic field, and does it ever flip North to South?.................... 109

Sure, we all know Earth has a magnetic field, otherwise compasses wouldn't work, but how is Earth's magnetic field generated? Is there a giant bar magnet within Earth? And does the strength of the magnetic field vary? Perhaps more importantly, has Earth's magnetic field ever dropped to zero (or close to zero) and would it matter if it did? Is the magnetic north pole (and magnetic south pole) fixed in one location, or does it move and if it does move, could our N and S magnetic poles ever flip, so that our N magnetic pole becomes our S magnetic pole? Does life on Earth benefit from the magnetic field around Earth? Has the magnetic field affected the evolution of humans? These questions form the basis of this chapter. But first, I include a point of clarification about the number of N and S poles that we have......

Chapter 7   What are viruses? Are viruses alive, and where did COVID-19 come from? 129

Viruses! We have heard more about viruses and inoculations in the past 36 months than we ever wanted. COVID-19, SARS (severe acute respiratory syndrome), Spanish flu, Ebola, Monkeypox, to name just a few that have been in the news. We know many seem to have a dislike for Humans. But what are viruses, and are they alive? How do we define 'living'? How do viruses infect us? Where did COVID-19 come from?

A truly rich tapestry of themes are discussed in this chapter.

Chapter 8 Size really does matter. What is the largest organism in the world? Why are the testicles of long-tailed Macaques soooo very large? 155

A dinosaur. A blue whale. A giant redwood tree. These are the most common contenders for the prize of the largest living or extinct organism in the world. But are any of these answers correct?

Humans are fascinated by big things. Having the tallest building in the world (currently the Burj Khalifa in Dubai at 829.8 metres (2,722 ft) tall) is a much-sought-after title of many cities. The largest ship ever built (the Seawise Giant; 261,000 tonnes, 458 m long) and the largest man-made hole (Bingham Canyon Mine, an open-cast copper mine located 28 miles southwest of Salt Lake City about 4 km in diameter and about 1.2 km deep and visible from space) are notable achievements in size. Such man-made records, however, change with time and technology. But what of the realm of the living world? What is the largest single organism, including extinct and extant (still living) organisms? Do differences in size among different species affect any aspects of the physiology and ecology of these species? Is there such a thing as eternal life, and why are the testicles of long-tailed Macaques sooo very large? In this chapter I look at these questions.

Chapter 9 How did lightning simultaneously kill 62 cattle in a field in 2005? What makes lightning, and how is lightning globally important? 181

Lightning – we have all seen it and perhaps many of us were a little scared when we were children, either of the lightning itself or the following thunder, or both. But even as adults few would know how lightning can kill dozens of cattle in a single field without physically striking them. And few would realise the global importance of lightning. In this chapter I address the three questions posed in the title. In doing so, I touch upon lightning, wildfires, ozone production, ball lightning and St. Elmo's Fire. A veritable smorgasbord of topics.

Chapter 10 Why is the sky blue and grass green and how have satellites changed our ability to monitor the world's surface? 199

As indicated in the Preface, this was the first science-based question (well, the first part about the sky) that I recall asking my Dad 55 years ago. He didn't have an answer, but we spent several hours looking through a set of Encyclopedias he had recently bought, to find an answer. It is undoubtedly his fault that I became a scientist!

In this chapter, I start with a consideration of the wavelengths (colours) of light emitted by our Sun, before discussing the absorption and scattering of photons by gas molecules and dust in the atmosphere. The reason for the sky being blue on Earth, but black when viewed from the moon is all about scattering of photons by an atmosphere. When examining the question about why grass is green, some weird similarities between the structure of haemoglobin in Human blood, chlorophyll in plant leaves, foetal haemoglobin in the blood of an unborn baby, and the nodules of roots of legumes (peas and beans, for example) are uncovered. We also discover that chlorophyll can be bright red, despite usually looking green. I then spend time talking about how the field of remote sensing from satellites uses this fact to measure the greenness of terrestrial landscapes and show some significant increases in terrestrial landscape "greenness" globally, and the importance of this greening. Finally, we see how global decreases in sea levels recorded in 2010/2011 were detected by a system measuring the 'breathing of landscapes' located in the middle of Australia. An absolute medley of topics to ponder.

Chapter 11 What is the difference between weather and climate? And which comes first, increased atmospheric CO_2 concentrations or increased surface temperatures? 229

The surface temperature of Earth is currently increasing. This is not a contentious statement – global warming is a reality. Similarly, global atmospheric CO_2 concentrations are increasing. That is also clear. However, which came/comes first – changes in mean surface temperature or changes in atmospheric CO_2 concentrations? If increases in temperature precede increases in atmospheric CO_2 concentration, is there a hypothesis that might explain how temperature increased prior to the Industrial Revolution? And, what has caused increases in atmospheric CO_2 observed after the start of the Industrial Revolution (about 250 years ago)? In answering these questions, I challenge the prevailing paradigm (with evidence), and [daringly] propose that our current response to global warming and climate change is not evidence-based, may be misguided, and the lack of robust cost-benefit analyses is leading us down the wrong response pathway.

Too confronting?

Chapter 12 What is Science and the scientific method? Why is the Science of climate change so NOT settled?.. 249

The phrase 'The Science is settled' is frequently used in the media (print and electronic) when discussing climate change. This means that no one is allowed to question the existence, and more importantly, the causes, of climate change, and no one is allowed to question the optimal strategies required in the future to either prevent further increases in temperature and atmospheric CO_2 concentration or learn to live with climate change. Because – **The Science Is Settled.........NOT.**

This chapter addresses four big questions. These are:

(a) What is Science and the scientific method?
(b) Can Science ever be settled?
(c) Is there evidence that human-induced increases in temperature are having a catastrophic impact on extremes, including drought, tropical storms (cyclones), the Southern Oscillation Index and the number and areal extent of wildfires?
(d) Is there an alternative to the net zero path currently being promulgated throughout the West?

That is certainly a comprehensive collection of contentious topics!

Curiosity Unleashed

Chapter *1*

What are jet trails and how are they made? What is the link between the 9/11 attack on the Twin Towers, jet trails, and land surface temperatures?

On the morning of the 9th of September 2001, four passenger airlines were hijacked by Islamic terrorists of the al-Qaeda group. One jet was deliberately flown into the North tower of the World Trade Centre (WTC) in New York, another was flown into the South tower of the WTC and a third jet was flown into the Pentagon in Virginia, USA. A fourth jet crashed into a field in Pennsylvania after a determined defence of the plane by passengers on board.

In response to the attack, all civilian flights into, out of, and across the USA were cancelled for three days. This was an unprecedented response to an unprecedented attack. But it provided a unique opportunity to test an important hypothesis: that contrails can alter the weather.

After looking at how contrails are made, and the importance of contrails to the surface energy balance of a landscape, this chapter then shows how the eruption of Mount Pinatubo in 1991 resulted in a massive increase in the rate of uptake of CO_2 into trees – so much so that atmospheric CO_2 levels stopped increasing for a year or two. The specific example of contrails and their impact on temperature and weather is merely one example of global dimming, whereby the amount of solar radiation arriving at the Earth's surface has declined for several decades. However global dimming has now been replaced by global brightening. How does global dimming and global brightening affect carbon uptake and water use by vegetation?

A common assumption made in the media and some scientific publications is that as the world warms up, the rate of evaporation from lakes, wet soil surfaces and vegetation must increase, in the same way that a hot pan of water on a cook top evaporates more rapidly than a cool pan of water sitting on a bench top. However, the Pan Paradox states that to the contrary of this widely held assumption, evaporation from open surfaces of water has been declining for the past 50 years, despite temperatures rising over that period. This paradox is discussed and put into the context of regional water budgets.

A veritable cornucopia of subjects to be addressed in this chapter.

CONTRAILS

Contrails (short for condensation trails) are clouds that form in the wake of jets, visible as long, narrow, straight lines of white cloud behind jets cruising at 30,000 feet (about 10,000 m; Fig. 1). They form because water vapour released from the jet engine as exhaust, along with some water vapour already in the atmosphere, freezes around small particles in the extreme cold (-35°C to -55°C) of the atmosphere at cruising altitude. Some of the small particles come from the engine itself (e.g., soot, metal filings) and some are already present in the atmosphere. Some contrails are short lived (< 5 min), appearing for only a short distance (and hence time) behind the jet. Some are long-lived, persisting more than an hour after the jet has passed.

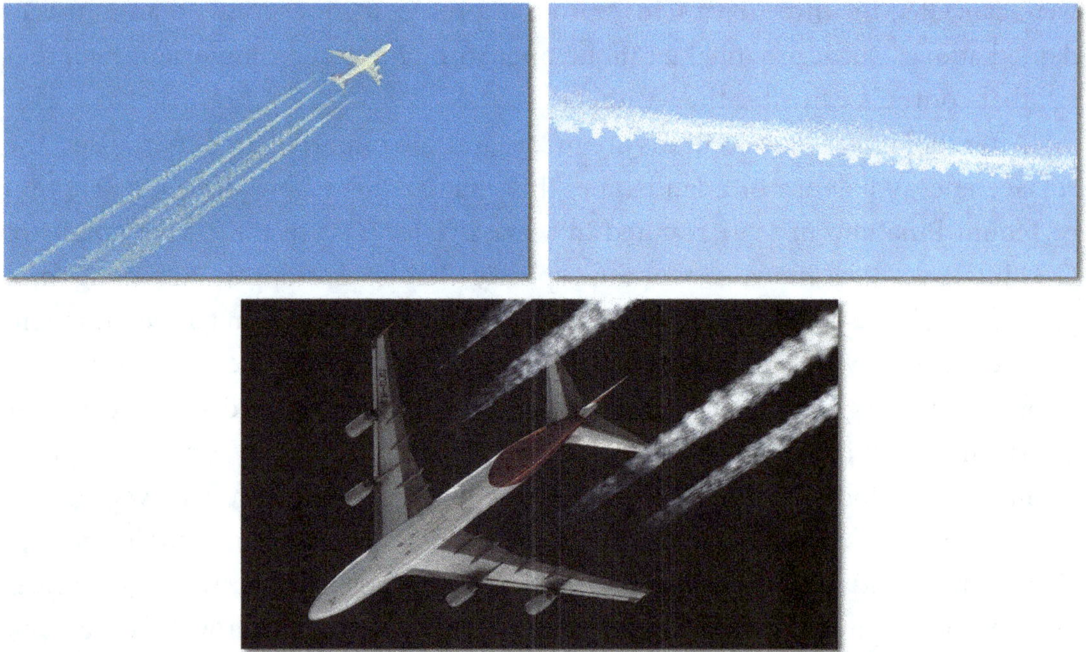

Figure 1:   Contrails (jet trails) are a common sight behind jets flying through clear blue skies. From: Wikipedia: https://commons.wikimedia.org/wiki/File:A340-313X.jpg

The more water vapour in the atmosphere, the longer the contrail persists. Some long-lived contrails remain as tight, non-spreading lines of cloud while others diffuse outwards and disperse, covering more of the sky and persisting for longer than non-dispersing contrails.

Contrails are always made of ice and only form at high altitudes (typically > 8000 m). In contrast, 'normal' clouds can contain ice and water vapour and can form at much lower altitudes (cloud at ground level is, of course, fog). A persistent contrail might contain 1 kg to 100 kg of water (as ice) per metre of contrail. In very busy jet traffic corridors (Fig. 2), the accumulation of spreading contrails from multiple jets can result in the formation of high-altitude cirrus, cirrocumulus, or cirrostratus clouds.

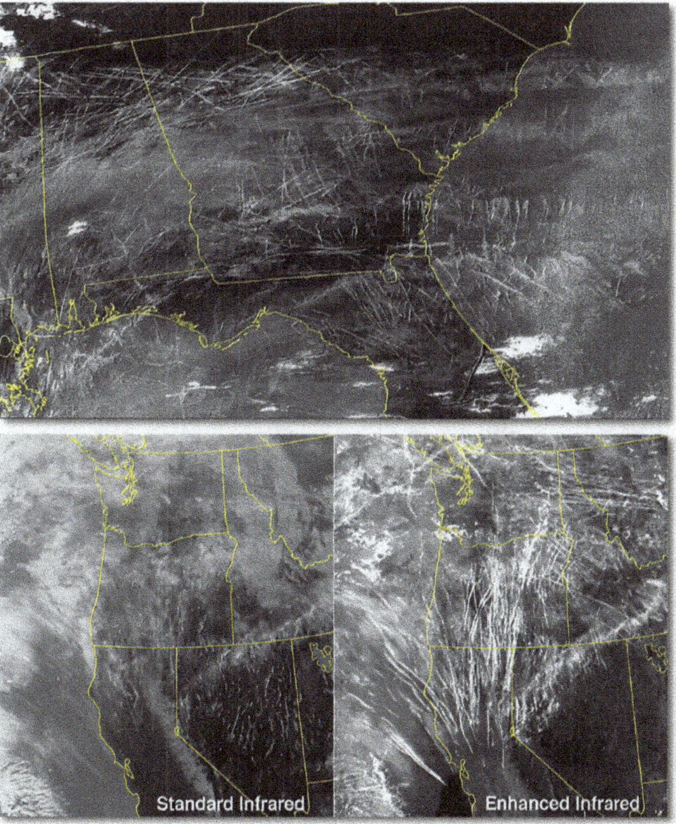

Figure 2: Images from the NOAA-17 satellite show how contrails can look like cloud cover. The upper and lower-left images are standard infrared satellite images, while the image on the lower-right uses two infrared wavelengths to reveal contrails. Images from: https://www.nasa.gov/centers/langley/news/releases/2004/04-140.html

Solar radiation and clouds

Our sun emits solar radiation in the waveband from about 100 nm to about 3000 nm, called short-wave radiation. This includes UV (wavelengths of about 100 – 400 nm; 1 nanometre (nm) is 1000th of a micron (μm) or 1,000,000,000th of a metre), infra-red radiation (700 nm to about 1 mm), and what we, as humans, see, the visible spectrum, 400 – 70 nm. Interestingly, this is the same waveband that plants use to drive photosynthesis. We can broadly define radiation at ground level into two classes, short wave (wavelengths 0.3 – 3 μm, or 300 – 3000 nm) and long wave (wavelengths > 3 μm) radiation. The surface temperature of an object determines the wavelength of peak emittance of radiation (i.e., light). The surface of the sun is very hot (about 5800 – 6000 K) and produces peak emittance at the short wavelength of 500 nm (which is the wavelength of blue light). In contrast, all objects at room temperature (let's say between 0 and 30 ^{o}C) emit longwave radiation. For example, ice, at 0 ^{o}C, emits radiation with a peak at 10,600 nm. A grass lawn, a rock in the desert, humans, trees, soil, all emit long wave radiation. The energy balance of the Earth's surface is the balance between incoming radiation (predominantly shortwave, but some longwave too) and the loss of energy as longwave radiation.

The amount of solar radiation arriving at the outer edge of the Earth's atmosphere above the equator is about 1.36 kW m^{-2} (or about 81.65 kJ m^{-2} per minute). This fluctuates by about 0.1 % according to where the sun is within its 11-y solar (sunspot) cycle.

The proportion of the total solar radiation emitted by the sun that reaches the *surface* of Earth is typically about 75 %. Short wave radiation is scattered, absorbed, or transmitted through the atmosphere. Scattering by molecules that are smaller than the wavelength of a photon of light is most pronounced for blue light, which is why the sky tends to look blue (Chapter 10), because blue photons are scattered towards the Earth's surface, while red photons are scattered less towards the Earth's surface. Ozone absorbs UV light, and diatomic gases such as CO_2, CH_4, and H_2O absorb infra-red wavelengths. Aerosols, clouds, dust, soot, and smoke can absorb short wave radiation and clouds are good at reflecting solar radiation (back out to space) and absorbing solar radiation, but also emit infra-red radiation. If all the clouds in the atmosphere disappeared for a day or two, the increase in solar radiation reaching the ground would be substantial because of the absence of the layer of reflective clouds.

Direct-beam radiation is the fraction of solar radiation that comes directly from the sun to the Earth's surface without being scattered by dust, aerosols, or clouds. Diffuse-beam radiation is the fraction of solar radiation that arrives at the Earth's surface after being scattered by aerosols, dust, and clouds. See below.

CONTRAILS AND LOCAL WEATHER

Clouds do complicated things to the Earth's energy balance (that is, the balance between incoming short wave solar radiation and outgoing long wave and reflected short wave radiation. See text box above).

Contrails (and clouds generally) can trap outgoing (emitted from Earth to space) long wave radiation emitted by everything on the Earth's surface, which means trapping heat, like blankets trap heat around us in bed. They can also reflect incoming solar radiation back out to space, thereby reducing the heating effect of incoming solar radiation. The relative effects of these two processes differ according to whether it is night or day.

During the day, reflection of solar radiation by contrails (and cirrus cloud formed by the convergence of contrails) is large, thereby reducing the heating effect of incoming solar radiation during daylight hours and reducing maximum daytime temperatures. During the night, when the sun isn't shining, reflection of solar radiation is absent, but the blanket effect of contrails is still large, so minimum night-time temperatures are slightly increased (because less long-wave radiation is lost to space so there is less cooling during the night). Consequently, the daily difference between maximum temperature (during daylight hours) and minimum temperature (at night), is predicted to be smaller when contrails and cirrus clouds are present than when they are absent. Testing this hypothesis was proving difficult until

David Travis and colleagues realised that the 3-day cessation of all flights across the USA following the 9/11 attack presented a unique opportunity to examine the hypothesis. By comparing daily maximum and minimum temperatures recorded at 4000 meteorological stations across the USA before, during, and after the 3-day ban on air traffic, Travis and co-workers established that there was indeed a 1.1°C increase in the diurnal temperature range arising from the cessation of flights above the USA following the 9/11 attacks, in accordance with the expected effect of contrails/cirrus cloud on land surface energy budgets.

There is also strong evidence that increased air traffic over the past several decades has increased the amount of cirrus cloud cover in regions where air traffic is especially heavy, for example in western Europe, the flight corridors over the Atlantic and Pacific oceans, and the major flight corridors of the USA. Consequently, the diurnal temperature range in such areas has declined significantly. Minnis and co-workers at NASA have shown that the increase in cirrus clouds arising from contrails should cause a tropospheric warming of 0.28 – 0.38 °C per decade, a range remarkably close to the observed tropospheric temperature rise of about 0.3 °C per decade between 1975 and 1994.

This effect of contrails on global energy balances is a specific case of a broader effect called global dimming, which I now discuss.

GLOBAL DIMMING

Global dimming is the reduction in the amount of sunlight (solar radiation) reaching the surface of Earth. Between the 1950's and the mid-1980's, there was an 8 % decline in the amount of solar radiation received by the Earth's surface in Antarctica, Europe (e.g., Fig. 3), Russia, the USA, and the Middle East, followed by an increase in solar radiation levels (solar brightening; Fig. 3). I discuss this recovery after I discuss solar dimming.

What might cause a decline in solar radiation levels received by the surface of Earth?

Global dimming (always recorded at ground level) is caused by increased absorption and reflection of solar radiation by aerosols and particles in the upper atmosphere. It is not related or caused by any changes in the amount of energy (solar radiation) emitted by the sun (but more of that later). It is a predominantly

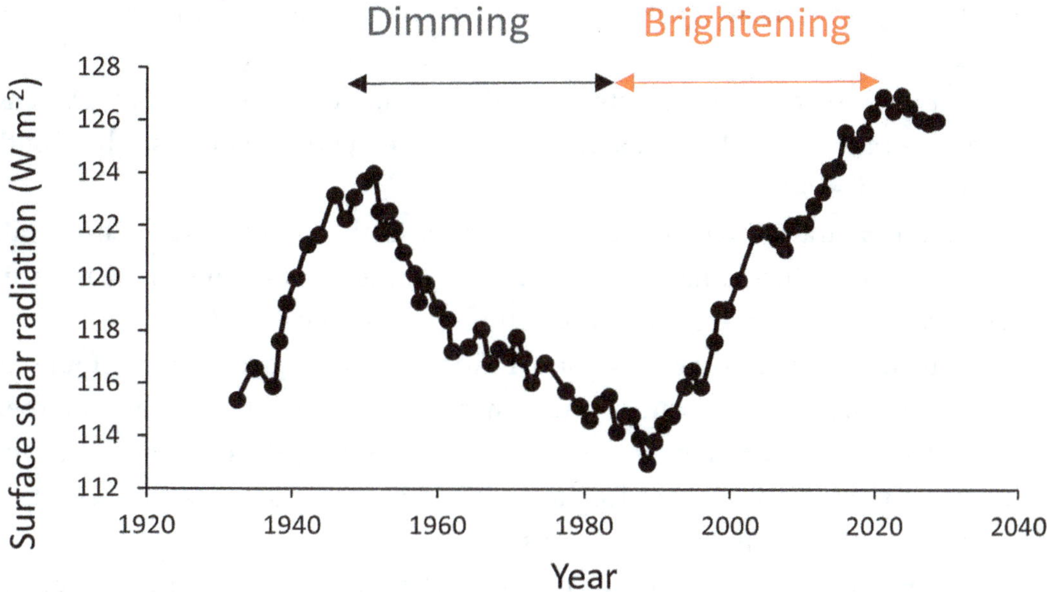

Figure 3: Surface solar radiation levels have fluctuated significantly over the past 70 years, reflecting periods of solar brightening and solar dimming. From https://iac.ethz.ch/group/climate-and-water-cycle/research/radiation-and-the-hydrological-cycle/global-dimming-and-brightening.html. Institute for Atmospheric and Climate Science.

anthropogenic effect, that is, it is caused by the activities of Humans, in particular the burning of coal, petrol, and diesel. There are also some natural sources of particulate matter in the upper and lower atmospheres. Volcanic eruptions release vast amounts of sulphur dioxide, ash, and smoke into the atmosphere. Sulphur dioxide is quickly converted to sulphate aerosols which, when ejected high enough (about 15 km in tropical zones), along with ash, can reflect significant amounts of incoming solar radiation, thereby cooling the environment for many months and even a few years. The particles from volcanic eruptions can circulate vast distances around the globe, exerting an impact far removed from the original source. For example, the eruption of Mount Pinatubo in 1991 in the Philippines ejected ash more than 30 km up into the atmosphere and cooled Earth by almost 0.5°C for two years (see below; Fig. 4). Volcanic eruptions can also affect large-scale weather patterns, including the location of the Intertropical Convergence Zone (ITCZ). The ITCZ is a zone at the thermal equator where trade winds converge and is a low-pressure belt that moves north and south of the geographical equator (0° latitude) seasonally. In the Austral summer, the thermal equator and ITCZ are south of the geographical equator, in the northern summer, the thermal equator and ITCZ are north of the geographical equator. This movement of the ITCZ is significant because it directly affects the trade winds, the distribution of warm and cold waters of the Pacific Ocean, and rainfall and temperature in Australia, South America, and elsewhere.

A second natural source of particulate matter in the atmosphere is large-scale vegetation burning, including the widespread burning of rangelands in California in 2016, which burnt more than half-a-million acres (almost 2300 km^2).

Particulate matter also acts as condensation nuclei, causing more clouds to form with smaller droplets which results in more solar radiation being absorbed and reflected, thereby enhancing the cooling effect during the day, with warming occurring at night, and thereby decreasing the diurnal range in temperatures, as noted above in the discussion of contrails.

So, global dimming is the result of accumulation of aerosols and particulate matter in the upper atmosphere, and these can arise from anthropogenic and natural causes. Since the mid-1990s, the particulate matter (aerosols) in the atmosphere has been gradually declining because of pollution controls in many countries, so the good news is that sulphate pollution is declining. However, the reductions in concentration of aerosols and particulate matter is associated with an increase in the amount of solar radiation reaching the ground, and this is leading to global brightening (Fig. 3). It has been suggested, though, that this global brightening will accelerate, or enhance, global warming.

Should we care about global dimming and brightening?

IMPACTS OF GLOBAL DIMMING/BRIGHTENING ON GLOBAL CARBON AND WATER CYCLES

When aerosols accumulate in the atmosphere, we see it as "smog". It might be a distinct layer in the atmosphere, or it might be a more diffuse "browning" of the atmosphere. Smog impacts human health, causing respiratory problems, especially for people with asthma. China is currently going through a particularly bad decade with respect to smog because of the rapid and poorly-regulated industrialisation of the country. But most industrialised countries have in the past, or are currently, suffering from the ill-effects of smog to a greater or lesser extent. The impost on human health is generally understood. What is not generally understood are the impacts on global carbon and water budgets (that is, the rate at which vegetation takes up/releases carbon and the rate at which vegetation uses water as transpiration from leaves). Let me explain.

Increased aerosol content of the atmosphere, and clouds (naturally-occurring or the results of contrails combining), cause more incoming solar radiation to be forward scattered (see Chapter 10). What we see is more diffuse beam radiation and less direct beam radiation arriving at the land surface, that is, less direct beam radiation and more diffuse beam radiation hitting plant canopies. When this happens, the volume of shade within vegetation canopies, but especially within the deep canopies of woodlands and forests, declines, sometimes by a factor of 10 or more. The reason for this is simple. When direct beam radiation illuminates a leaf, the leaf beneath is in shadow. The amount of light available to the lower leaf is reduced compared to the leaf above it. In contrast, when diffuse radiation hits the upper leaf, very little shadow is created because there is light arriving from the side and from all angles; hence the lower leaf is illuminated as much as the upper leaf. The volume of shade is reduced when more diffuse beam radiation arrives at the top of a canopy. We are all aware of this on a personal level. On bright sunny days we see lots of shadows on the ground as we walk outside. We even have one following us around, at our feet. But on overcast days when clouds obscure the sun, and all the light we see is made up of diffuse beam radiation, we don't see shadows on the ground because all objects are being illuminated equally on all sides – so no shadow!

Increasing the availability of light to all those leaves in the lower canopy means that the rate of photosynthesis of the lower leaf is larger under diffuse beam conditions (i.e., on cloudy and smoggy days) than under direct beam conditions (on clear blue skies). Most leaves in a woodland or forest are shaded within the canopy. Very few receive direct sunlight. Therefore, the stimulation of photosynthesis of all the leaves in the lower canopy is substantial on cloudy or smoggy days. In fact, it is so large that it can be detected at large-scales.

Rap and his many co-workers have shown that when farmers burn vast areas of the Amazon to clear the forest for farming, the aerosols produced increase the diffuse beam fraction by about 5 % and as a result, the net primary productivity (the net gain in carbon (C) stored in the vegetation) of the Amazon increases by about 2 %, or about 100 Tg C per year, or 100 million tonnes of C per year. Similar stimulation of C uptake on cloudy days has been observed in China and the eastern USA. Truly, this is a global phenomenon of massive proportions.

As mentioned above, when volcanoes explode, they eject huge quantities of ash and sulphur dioxide, which rapidly produce aerosols in the upper atmosphere. These aerosols can persist for many years and travel around the world, borne by winds circulating the globe at high altitude. The 1991 eruption of Mt Pinatubo ejected about 10 billion tonnes (or 10 km^3) of magma, and 20 million tonnes of SO_2 into the stratosphere, resulting in the formation of a world-wide layer of sulphuric acid haze. The increase in atmospheric haze (aerosols) greatly increased the diffuse beam fraction of solar radiation for vast areas of the globe. Global temperatures dropped by about 0.5 °C (0.9 °F) and atmospheric CO_2 levels declined by about 2.0 ppm between 1991 and 1993 (Fig. 4). One suggested cause of this decline in atmospheric CO_2 levels was the stimulation of photosynthesis of the world's woodlands and forests arising from the increase in diffuse beam solar radiation arising from the aerosols that were ejected into the atmosphere by the eruption. This decline in atmospheric CO_2 levels represents an increase in C uptake of about 2.75 Gt of C or 2750 million tonnes of C. Note that globally we are emitting about 10,000 million

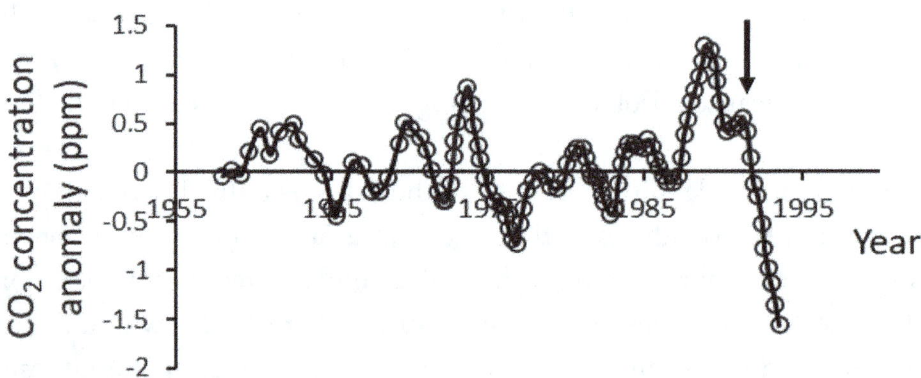

Figure 4: The CO_2 concentration anomaly, calculated by removing the seasonal signal (the seasonal oscillation arising from the balance between photosynthesis and respiration of vegetation) and detrending the remaining values of concentration by assuming a constant fraction of annual industrial emissions of CO_2. El Nino events generally reduce global rates of CO_2 uptake and so small increases in atmospheric CO_2 concentration occur, but the Pinatubo eruption (denoted by the black arrow at June 1991) caused a large increase in global CO_2 uptake because of the increase in diffuse beam solar radiation. Redrawn from Sarmiento 1993 Nature, vol 365, p693-694.

tonnes of C per year from fossil fuels so that the increase in CO_2 uptake resulting from one volcano eruption is more than one quarter of global annual emissions of C. I describe two other possible causes of the decline in atmospheric CO_2 concentrations from 1991 to 1993, below.

The ability of aerosols and clouds to reflect incoming solar radiation, and their impact on diffuse beam radiation and hence photosynthetic uptake of CO_2 have led to serious suggestions to globally geoengineer the upper atmosphere by releasing vast amounts of sulphur dioxide into it, to produce aerosols. Essentially the suggestion is to mimic the effect of volcanic eruptions on global temperature and global photosynthetic C uptake. This would occur through two mechanisms. First, it would increase the diffuse beam fraction of solar radiation. This will increase the rate of photosynthetic C uptake by trees globally, reducing atmospheric CO_2 concentrations. Second, it will reflect more solar radiation back to space, thereby reducing global warming. Xia and his colleagues (Xia et al., 2016, Atmospheric Chemistry and Physics) estimate that 8 million tonnes of SO_2 could increase annual global C uptake by almost 4000 million tonnes, which would greatly reduce the rate of increase in atmospheric CO_2 concentrations.

ALTERNATIVE EXPLANATIONS OF THE DECLINE IN ATMOSPHERIC CO_2 CONCENTRATIONS

Two alternative explanations for the decline in atmospheric CO_2 concentrations after the eruption of Mount Pinatubo have been made. Angert and co-workers (Angert et al., 2004, Geophysical Research Letters) used two models to predict changes in atmospheric CO_2 concentration and concluded that increased diffuse beam solar radiation levels could not explain the observed decline in atmospheric CO_2 concentration. They also rejected a role for a decline in respiration of soils and forests following the eruption. Soil and vegetation respiration releases very large amounts of CO_2 into the atmosphere. Because respiration is more sensitive to temperature than photosynthesis, the observed decline in global mean temperatures would have undoubtedly reduced the release of CO_2 from respiration. If this reduction was proportionally larger than any reduction in photosynthesis arising from reduced temperature, the rise in atmospheric CO_2 levels would have been reversed (for a short while). They concluded that a global reduction in biomass burning (which releases CO_2 to the atmosphere) and an increase in oceanic uptake of CO_2 caused the temporary decline in atmospheric CO_2 concentrations following the eruption. I am unaware of a reason for any increase in oceanic uptake of CO_2 following the eruption and I am convinced that increased diffuse beam radiation and the resulting increase in vegetation uptake of CO_2 is the correct explanation.

THE WATER CYCLE, PAN EVAPORATION, SURFACE ENERGY BALANCES, GLOBAL BRIGHTENING, AND THE PAN PARADOX

Global brightening has been recorded since the mid-1980's. Any increase in solar radiation inputs to the atmosphere are associated with an increase in near-surface atmospheric temperature, irrespective of whether atmospheric CO_2 levels increased. The water and carbon cycles are linked because of the central role of vegetation in both (see text box below). Having talked about CO_2 uptake by vegetation earlier in this chapter, I now want to focus on the water cycle, and particularly on evaporation of water from 'open-water bodies' (such as lakes, oceans, a large pan of water sitting in a field). Evaporation from an open-water body is conceptually similar to evaporation of water from inside a leaf to the air surrounding a leaf. The major difference, however, is that leaves can close the pores (stomatal pores; see text box) through which transpiration (evaporation of water from inside a leaf to the atmosphere around the leaf) occurs; open bodies of water can't do that.

Pans of water have been used to measure evaporation to the atmosphere for more than two centuries (since the mid-17th century). They were, for many decades, used to assist farmers and horticulturalists on the timing of crop irrigation. When a large volume of water has evaporated into the atmosphere from the pan (as indicated by a drop in depth of water in the pan), it is probably wise to irrigate your crop. Figure 6 shows a typical Class A evaporation pan. It must be level and not subject to shadows from vegetation or buildings. A wire frame is used to stop animals drinking the water and rain gauges are required to measure any input of rainwater.

The link between CO_2 and water fluxes

Plants absorb CO_2 during photosynthesis but they release water vapour at the same time in a process called transpiration. The uptake of CO_2 during photosynthesis and transpirational loss of water occur *almost* exclusively via pores in leaves called stomata (singular, stomate). Stomata in trees and most crops open during the day to allow uptake of CO_2, with sunlight providing the energy required to enzymatically "fix" CO_2 into sugars (thereby driving the production of plant products, including, for example, wood, cotton, and the food we eat). Because the concentration of water vapour inside the leaf exceeds that of the water vapour in the air around the leaf, water molecules diffuse out through the open pore. When soil moisture levels and solar radiation levels are moderate-to-high, stomata are fully open and the rate of photosynthetic uptake of CO_2 is maximal. Under these conditions, transpiration rates are also high, requiring rapid movement of water from the soil to the canopy, via roots and stems.

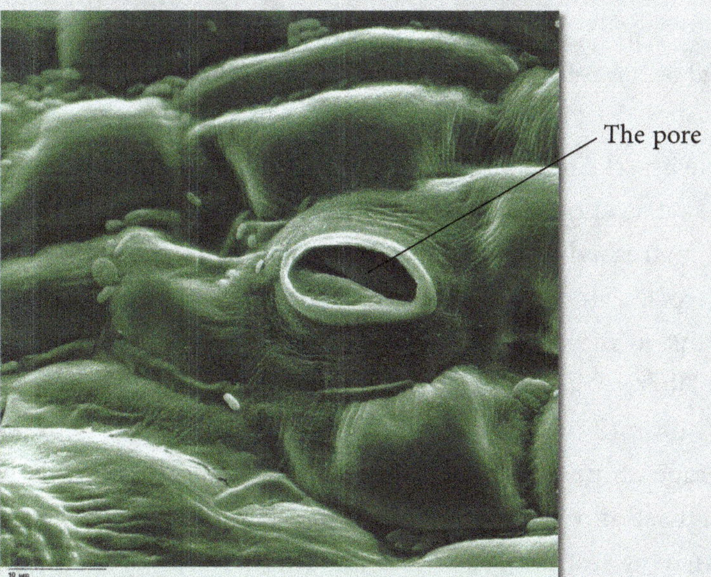

Figure 5: A surface image of a tomato leaf showing a single stomatal pore (mostly closed). Open pores allow CO_2 to diffuse into the leaf and water vapour to diffuse out. The pore is approximately 12.5 µm wide. Image from https://en.wikipedia.org/wiki/Stoma#/media/File:Tomato_leaf_stomate_1-color.jpg

Figure 6: A Class *A* Pan, with an associated anemometer. From wiki: pan evaporation. Image created by Bidgee under Creative Commons Attribution 3.0 Unported License.

Most people assert that a globally warmer temperature must be associated with increased rates of evaporation of water from wet surfaces into the atmosphere. Consequently, increased rates of Pan Evaporation (P_aE) are predicted to occur in a warming world. However, declines in P_aE have been recorded from multiple evaporation pans, at multiple sites, across many recent decades. Studies in China, Australia, the Tibetan Plateau, Thailand, India, and the USA have recorded rates of decline in evaporation of between 0.7 mm y^{-2} and 12 mm y^{-2} between 1948 and 2005. The average annual decline in evaporation rate during this period, including the two highest annual rates of 12 mm y^{-2} and 10.5 mm y^{-2}, is 3.85 mm y^{-2}; excluding the two highest rates as potential outliers yields an average of 2.51 mm y^{-2}. (If you are wondering why the units are mm per year per year (mm y^{-2}), it is because it is the *rate of change* of annual rates of evaporation, it is not the annual rate of evaporation.) An average decline of 2.51 mm y^{-2} might not seem much, but it is very significant in energy terms. Let me explain. The latent heat of vapourisation of water (this is the amount of energy that must be absorbed by liquid water to convert it to water vapour, i.e., to vapourise the water) is 2.45 MJ kg^{-1}, so it takes 2.45 MJ of energy to vapourise 1 kg (1 litre) of water. A 1 mm-depth of water evaporated from 1 m^2 of surface is equal to 1 kg of water evaporated from that surface, which requires 2.45 MJ of energy (equivalent to 2.45 MW s^{-1} because 1 W = 1 J s^{-1}). An evaporation rate of 2.51 mm y^{-2} is therefore equivalent to 6.15 (i.e., 2.45 x 2.51) divided by the number of seconds in 1 year (=31536000 s) = 0.195 MW m^{-2} y^{-1} equivalent to 1.95 W m^{-2} over a 10-year period. Earth, including its atmosphere, has an energy imbalance of between 0.4 – 1 W m^{-2}. The energy imbalance is the difference between the amount of solar radiation arriving at the top of Earth's atmosphere and the amount of radiation emitted back out to space by the Earth. Because the energy imbalance is positive, this means Earth is warming up (retaining heat). Over a 10-year period, the average decline in rates of evaporation of water from continually wet surfaces (and remember, evaporation of water absorbs heat from the atmosphere) is a significant fraction of the global energy imbalance.

What I have just described is known as the "**Pan Paradox**".

The Pan Paradox encompasses the following phenomena:

1. Global mean surface temperatures have increased significantly during the 20th century (Fig. 7).
2. A warming trend is expected by many people to result in increasing rates of evaporation from a water surface to the atmosphere.
3. However, contrary to expectation, P_aE rates have declined for the past 50+ years.

This is an uncontested and global phenomenon.

The paradox is: how can a warming temperature trend be associated with a declining rate of evaporation?

What determines the rate of evaporation of water from a water surface? There are several simple physical factors affecting the rate of evaporation, namely wind speed, humidity, temperature, and solar radiation. It turns out that the contribution of reduced radiation input arising from global dimming was relatively small (but significant) when we were experiencing dimming, but we are now (probably) in a

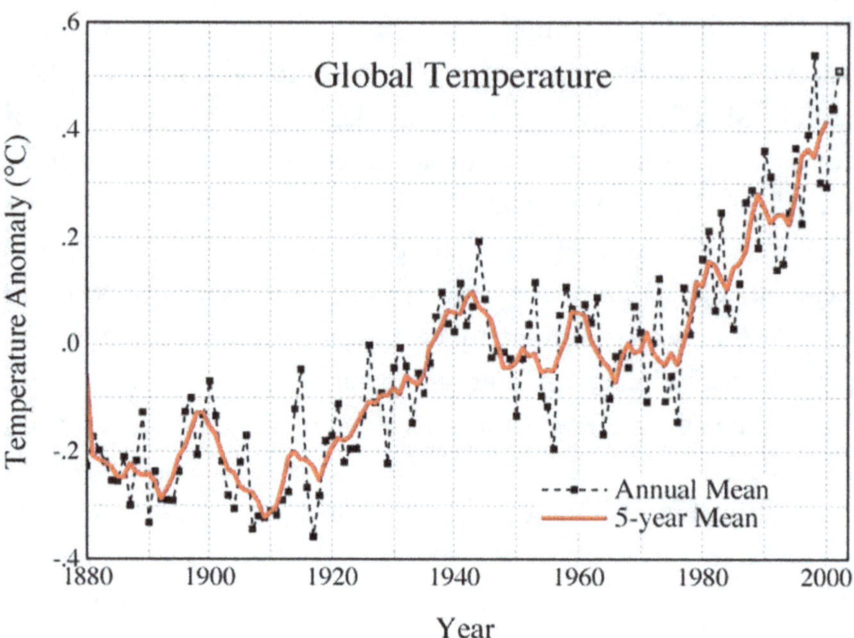

Figure 7: Global average surface temperature trends of the past 100+ years. From NASA: https://data.giss.nasa.gov/gistemp/2002/

phase of global brightening (some regions still appear to be experiencing dimming, possibly because of local atmospheric pollution issues). Global brightening can't cause a reduction in rates of P_aE rates because increasing land surface temperatures don't reduce rates of P_aE rates, and relative humidity remains approximately constant, despite temperatures increasing. Thus, the largest single factor contributing to the Pan Paradox is **global stilling**, that is, a reduction in wind speed, which I will discuss shortly.

Before discussing what might be causing a reduction of average wind speeds globally, I need to clarify one important point about the interpretation of reduced rates of P_aE, an interpretation that is often made, incorrectly, in the media and even in some scientific literature.

It is sometimes stated that P_aE rates reflect rates of vegetation water-use (that is, transpiration). This is because the insides of leaves are fully hydrated, and they lose water through stomata in a process that is equivalent to evaporation of water from the water surface in a Class A pan. Indeed, transpiration from leaves is influenced by wind speed, temperature, humidity, and solar radiation, just like evaporation from pans of water. This is true for the washing we hang out to dry on a washing line. Consequently, it has been concluded (incorrectly) that a reduction in P_aE will be mirrored by a reduction in rates of transpiration (that is, rates of vegetation water-use). However, P_aE rates are, in fact, a measure only of the evaporative demand of the atmosphere, essentially a measure of the capacity of the atmosphere to "extract water" from a wet surface. In contrast, the rate of transpiration from a leaf, or the rate of water use by a canopy of leaves, is determined by atmospheric demand **and the rate of supply of water** to the canopy from roots. The rate of supply of water from roots is primarily determined by the water content of the soil, especially in the upper 1 metre of soil, and this is mostly determined by local rainfall. But the situation is even more complex than this. Some sites are water-limited, which means that annual rainfall is much lower than what the atmosphere could evaporate in a year if more water was available (in the soil) for evaporation. Deserts are water-limited environments. All the rainfall that falls evaporates very quickly and the sparse vegetation has evolved to conserve water (for example, cacti store water in their stems and have no leaves). Runoff of water does not occur, and rivers do not generally flow out from such areas. Other locations are light-limited (also called solar radiation-limited). These sites are energy-limited. The amount of annual rainfall exceeds the energy available (solar radiation) to evaporate all the rainfall in one year, so runoff of water is generated, rivers flow, and groundwater is replenished. Rainforests are solar radiation (energy-limited), not water-limited, in terms of their productivity and rates of water use, and vegetation is abundant.

Therefore, multiple combinations of changes to P_aE rates, rainfall, and water-limited *versus* energy-limited sites are possible, and this combination needs to be known locally to make any informed estimate of what rates of vegetation water use will do at sites experiencing changes in P_aE. For example, in a rainforest (which is radiation-limited), if P_aE declines and rainfall doesn't change, vegetation water use is likely to decline. In a semi-arid woodland rainfall is much less than the amount of evaporation that could occur if more water was available, and rates of evaporation are limited by the supply of water (rainfall), thus if P_aE declines, the rate of vegetation water use *could* increase *if* rainfall increased (because there would be a more extensive vegetation canopy growing there because of the increase in rainfall). In contrast, vegetation water use could decline when P_aE declines if rainfall remains the same or declines. Thus, predicting vegetation water use from changes in P_aE is not as simple as some commentators assume.

GLOBAL STILLING

But back to global stilling……

Surface wind speeds have been declining in China, the USA, Australia, and much of Europe and Russia for several decades (50+ years). Annual mean wind speeds have been declining by about 0.1 to 0.15 m s^{-1} decade^{-1}, or about 10 % (perhaps up to 20 % in some regions) in 30 years. A decline of about 0.1 m s^{-1} doesn't sound much but remember that average wind speeds are quite small to begin with (about 2.5 m s^{-1}). In addition, high wind speeds appear to have declined more than low wind speeds. What might be causing the decline in average wind speeds?

Changes in patterns of atmospheric circulation likely account for 10–50 % of the decline in surface wind speeds and simulations using mesoscale models suggest that an increase in surface roughness (see below) could account for 25 – 60 % of the stilling. Most importantly, regions exhibiting the largest stilling generally coincide with regions where the amount of biomass has increased over the past 30 years, supporting the view that increased vegetation cover, especially woodlands and forests, has increased surface roughness and hence caused a reduction in wind speeds. But what is surface roughness?

Surface roughness is an attribute of a surface that determines how much the surface interacts with turbulence in wind as the wind flows over it. The easiest way to think about it is through a simple diagram (Fig. 8). In Figure 8, a small parcel of air (represented by a blue circle) is travelling above one of three surfaces. In Figure 8A the surface is very smooth and exerts very little drag on the movement of the air parcel, so its speed is high, and it travels easily across the surface. As surface roughness increases, as illustrated in Figure 8B and then Figure 8C, the drag exerted on the surface increases and the speed of the air parcel/wind declines. In the real world, the surface of a mown lawn, for example, has a low surface roughness — or resistance to flow — and exerts minimal drag on a wind travelling over it. The surface of a field of wheat, however, has a 'rougher surface' and exerts a larger drag, while the surface of a savanna, with 12 m tall trees spaced 10 m apart and an understorey of tall grasses and woody shrubs, has an even larger surface roughness and exerts a larger drag on any wind passing over it. Indeed, the savanna "captures" some eddies of wind as they pass over and "pulls" them through the canopy of the trees into the lower reaches of the grass and shrub layers.

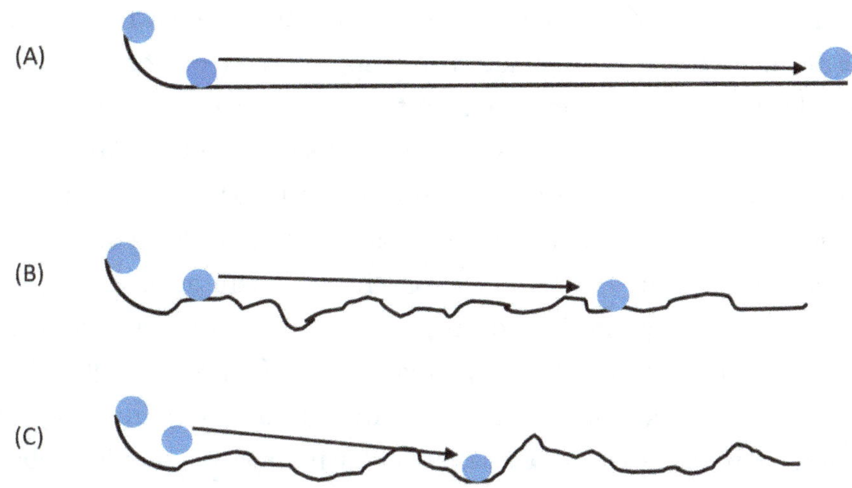

Figure 8: A crude representation of surface roughness and its impact on the flow of wind over the surface, and hence wind speed. The slope at the left end of each of (A) – (C) is used to give the ball (the blue circle at the top of the slope in (A-C)) some momentum once it is released. The blue ball represents a parcel of air in a wind. In (A), the surface is very smooth, it has a very low surface roughness. The ball can travel a substantial distance to the right, within 2 seconds (for example), so the average speed of the parcel of air (the ball) over the surface is high. In (B), the surface is much rougher, and travels a smaller distance in 2 seconds than in (A) so its average speed is lower. In (C) the surface is very rough, the parcel of air travels a smaller distance than either (A) or (B) and so its average speed is the smallest.

So why has land surface roughness increased in the past 50+ years? According to the Global Forest Assessment Report of 2005, the area of privately owned forest globally has increased by almost 2.7 million hectares per year, or 34 % overall, between 1990 and 2000. This may surprise many people, but this trend has been observed in many regional and global forest studies. For example, Liu and co-workers (2015, Remote Sensing) used remotely sensed data (reflectance values of a surface, whether that is a piece of land in the Amazon Rainforest, a savanna site in northern Australia, or a crop field in Europe) and established that, on average, the world became greener (more vegetation cover) between 1982 and 2012 (they used data only up to 2012), especially in India, Europe, North America, northern Southern America, and equatorial sub-Saharan Africa. Similarly, in 2022, Rifai and co-workers (2022, BioGeoSciences) demonstrated that across Australia, despite increasing dryness and increasing temperature over the past 38 years, the Australian landscape has "greened" significantly. This increase is associated with a decline in runoff of water across Australia's drainage basins, as must happen if there is more vegetation (because the vegetation transpires water out of soil and into the atmosphere, leaving less water to runoff to the sea in rivers).

There is no doubt that Earth has experienced significant greening for the past 40 years. Reasons for this include a warming Earth (especially important for cool and cold regions, where low temperature is the principal limiting factor for photosynthesis and growth), the CO_2 fertilisation effects (fertilisation here refers to the enhanced rate of photosynthesis and growth that occurs in response to increased atmospheric concentrations of CO_2), increased use of irrigation, global brightening, and changes in vegetation management. Increased greening is associated with increased surface roughness and hence reduced wind speeds, that is, global stilling. QED!

WHY DO WE CARE ABOUT PAN EVAPORATION RATES AND GLOBAL BRIGHTENING AND GLOBAL STILLING?

Understanding regional water budgets, including changes in river flow, soil moisture content, and groundwater recharge (Chapter 5), is very important to farmers, water supply companies, land and water resource managers, engineers, and many others.

As I previously discussed, at energy-limited sites, declining P_aE usually means a reduced rate of vegetation water use. Assuming rainfall does not change, some combination of runoff to rivers, soil moisture content and groundwater recharge, will increase in response to a declining P_aE, and this has been demonstrated in numerous studies around the world.

The impact of global dimming and brightening is apparent in global trends in rainfall anomalies (Fig. 9). Rainfall anomalies (the difference between the observed and a long-term average value) declined between the 1950s and the early 1980s, when global dimming was observed, but increased thereafter, correlating with global brightening. Changes in rainfall do, of course, have very large impacts on water available for drinking, irrigation, groundwater recharge, and ecological health.

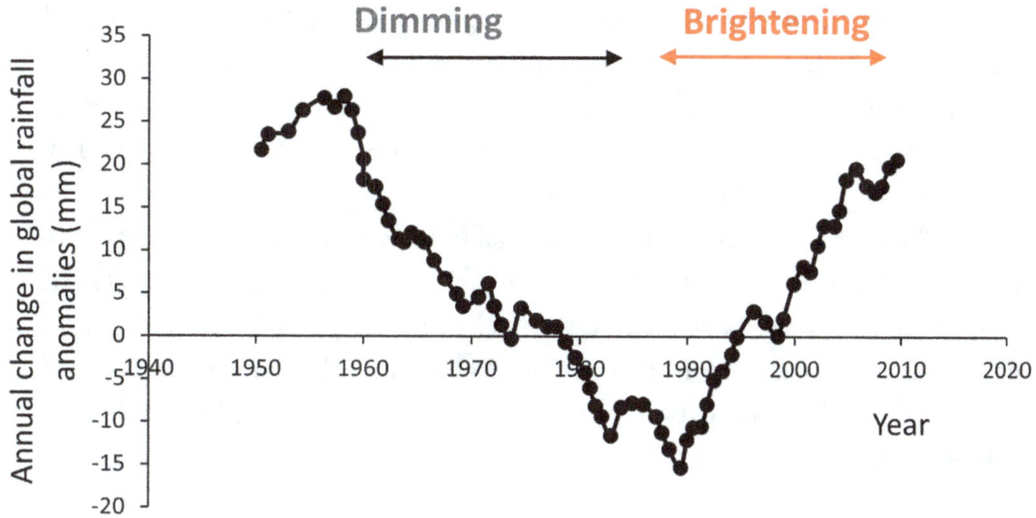

Figure 9:   Changes in global annual anomalies of average rainfall, illustrating a potential link between global dimming/brightening and rainfall. Redrawn from Wild, 2012, Bull. Amer. Met. Society.

Might global stilling reduce electricity production from wind turbines? That is a question for future debate.

To summarise, this chapter discussed the following key points:

Contrails, those fluffy white trails behind jets flying at 10,000 m, form when water vapour released through the combustion of jet fuel, freezes around particles in the extreme cold of the upper atmosphere. Contrails are therefore composed of ice. After the 2001 September 9th attacks on the World Trade Centre and the Pentagon in New York, all civilian flights over the USA were cancelled for 3 days. This meant that no contrails were produced for 3 days. This profoundly altered the radiation balance of the land surface of the USA on those days, resulting in a 1.1 °C increase in the diurnal temperature range (the difference between night-time minimum and daytime maximum temperatures). The effect of contrails on global energy balance is a particular instance of global dimming, whereby the amount of solar radiation reaching the Earth's surface declines. Extended periods of global dimming and global brightening have been observed during the 20th century. Both global dimming and brightening have affected global carbon and water cycles, with implications for food production, ecosystem function (for example, rates of carbon uptake and rates of transpiration). Volcanoes eject vast quantities of ash and sulphur dioxide and this can affect global energy balances for weeks, months, or years.

The Pan Paradox can be summarised as follows: while global average surface temperatures have been increasing for the past century or so, rates of evaporation of water have declined over the past 50 years, instead of rising, as would be expected from a consideration only of the increase in surface temperature. This could be explained by decreased average wind speeds (global stilling).

THE END

CHAPTER 2

WHY DOESN'T THE MOON FALL TO EARTH? AN EXPLORATION OF GRAVITY, TIDES, JELLYFISH, AND THE MOON (AND WHY WE NEVER SEE BOTH SIDES OF THE MOON, DESPITE THE MOON SPINNING ON ITS AXIS).

We all know the Moon makes the tides on Earth, right? But the sun is 27 million times heavier than the Moon and only 390 times further away – so does the sun have any influence on our tides? And why do we have two high and two low tides per day but only one Moon? How do jellyfish come into all this? And what can satellites measuring small differences in the gravity field around Earth tell us about water in and on Earth? This chapter tackles these issues and reveals how new satellites are being used to look at changes in the Earth's gravity and water supplies. This chapter also addresses the questions: 'Why doesn't the Moon fall to Earth – and what would happen if it did?' Oh yes, finally, as an aside, I answer the question: If the Moon is spinning, why do only ever see the same side whenever we look up at the Moon?

WHAT MAKES TIDES?

Gravity is the simple answer. But gravity from what?

The Moon weighs about 73×10^{18} tonnes (73 followed by 18 zeros) and is about 384,835 km from Earth. The Sun weighs about 2×10^{27} tonnes and is about 149,785,000 km from Earth. So, the Sun is 27 million times heavier than the Moon and only 390 times further away. The tidal generating force (gravitational pull) of the Moon or the Sun is proportional to the mass divided by the cube of the distance (that is, the distance multiplied by the distance multiplied by the distance) from Earth. If you do the maths, you find that the Sun has almost half (46 %) of the pulling power of the Moon – still a sizable effect! Newton realised this about 330 years ago. So, the Moon dominates in the formation of tides, but the Sun also exerts an influence, but in more subtle ways.

Is this why there are two high and two low tides each day – a "Moon" and a "Sun" tide? No. Inertia (the tendency for a mass to resist change in speed or direction of travel), plus gravity, explain the two tides per day. The gravitational pull of the Moon is largest when Earth is closest to the Moon (because of that cubed distance rule; the smaller the distance the larger the pull). The pull of the Moon is sufficient to overcome the inertia of oceans located on the side of Earth closest to the Moon. This means these oceans bulge upwards towards the Moon, creating the first high tide of the day for that part of the world. On the opposite side of Earth, inertia (or, as I prefer, the momentum) of the oceans exceeds the gravitational pull of the Moon (because this ocean is further from the Moon than the ocean on the side of Earth facing the Moon) Earth facing the Moon), and these oceans try to keep moving in a straight line, (because Earth, and hence the ocean, is rotating at a speed of about 460 m per second). Any rotating object (including oceans) tends to try and shoot off at a tangent to the circular motion. The ocean doesn't leave Earth on a tangent into deep space because of Earth's gravity, but it does create a bulge of water on the side of Earth furthest from the Moon, resulting in the first high tide of the day for that side of Earth. Of course, 12 hours later Earth has rotated through 180 degrees. Now the relative positions of the two (east and west) hemispheres of Earth have swapped. The hemisphere that was furthest from the Moon is now closest to the Moon and its ocean is pulled towards the Moon, creating its second tide, while the hemisphere that was closest is now furthest and momentum effects create the second high tide of the day for that ocean. These two bulges remain aligned with the Moon because water is a fluid not a solid and so can flow. So, two high tides and two low tides (which occur about 6 h after each high tide) per day occur across much of Earth. In fact, it is the lunar day that determines the timing of the tides and a lunar day is 24 hours and 50 minutes. The reason this is not 24 hours is because the Moon is rotating around Earth in the same direction that Earth is rotating and it takes an extra 50 minutes for Earth to catch up with the Moon.

The Sun's effect on tides is mostly on the size of the tide rather than creating the tide *per se*. Every lunar month there are two Spring tides and two Neap tides.

When the Sun and the Moon are aligned *on the same side of Earth*, the high tide is largest (called a spring tide) because the Sun and Moon are pulling the oceans in the same direction. One week later the Sun and the Moon are at right angles (90°) to each other, and tide heights are the smallest (called neap tides) because the Sun is pulling the ocean at 90 degrees to the direction that the Moon is pulling the ocean, and *vice versa*. There is no additive effect between the gravity of the Moon and the gravity of the Sun. One week later again, and Earth is between the Moon and the Sun, which are aligned again, but with the Moon and Sun on opposite sides of Earth. We have a second spring tide because the gravitational pulls of the Moon and Sun on oceans on opposite sides of Earth are maximal because of the absence of a 90° angle between Earth and Sun. Also, *the difference* in gravitational pull of the Moon on oceans on the side of Earth facing the Moon, and the gravitational pull of the Moon on oceans on the side of Earth facing the Sun is largest now (so the Moon's gravity is having a smaller impact on oceans on the side of Earth facing the Sun), so the Sun exerts its largest pull on the oceans facing the Sun when the Moon is on the opposite side of Earth. Hence the second Spring tide.

The path of the Moon is elliptical around Earth and therefore the distance between Earth and the Moon varies from a minimum of about 363,296 km to a maximum of about 405,504 km. Remember that the tidal pull of the Moon is smaller when the Moon is furthest away, and largest when the Moon is closest to Earth. The impact of the Sun on tides also varies as the distance from the Sun increases (moving from summer to winter) and decreases (moving from winter to summer) throughout the year. Because the distance between the Sun-Earth and Moon-Earth are not constant, the tidal range (the difference between the height of adjacent high and low tides) is not constant but changes in a predictable way.

Throughout this discussion I have assumed two low and two high tides every lunar day. A lunar day, as indicated above, is 24 hours and 50 minutes (the time taken for an observer on Earth under a specific position of the Moon on day 1, to get under the same point on the Moon's surface after one lunar rotation). High and low tides are therefore 12 hours and 25 minutes apart, for most places, such that, *for most places*, two high tides and two low tides occur every lunar day (that is, every 24 hours and 50 minutes).

JELLYFISH ARE AS IMPORTANT AS TIDES IN MIXING OCEANIC WATER

Upwelling is an important process in the world's oceans. Upwelling brings cold, but most importantly, nutrient-rich water from deep layers of the oceans to the surface. This nutrient-rich water stimulates and supports the growth of phytoplankton (single-celled photosynthetic organisms), which in turn supports large numbers of zooplankton (microscopic-to-small animals that can't swim or can only weakly swim, in oceans, rivers, and lakes), and hence fish, which eat the zooplankton. There are five major coastal regions where upwelling is pronounced, including the California current off the south-west coast of the USA and the Humboldt current (also known as the Peru current) off the coast of Peru and Chile. All five of these coastal sites of substantial upwelling support major fishing industries. Winds, in combination with the Coriolis effect and the Ekman transport effect (see Text Box), generate upwelling and the importance of upwelling to oceanography, marine biology, and even meteorology, is firmly established. But recently, a possible role for jellyfish in the mixing of deep and shallow ocean waters has been examined.

The Coriolis effect and Ekman transport: *it is all rather complicated*
Cyclones rotate in an anti-clockwise direction in the northern hemisphere but in a clockwise direction in the southern hemisphere. This is because Earth is rotating eastwards (i.e., from west to east), or anti-clockwise when viewed from the north polar star, and the opposite directions of rotation of cyclones are explained by the Coriolis effect. Imagine being a gunner and firing a massive artillery shell from a battleship located on the equator. You fire the massive shell directly north to a fixed target, 35 km away. Because you are travelling faster than the target (because the equator rotates at a higher speed than the target 35 km north of the equator), the shell lands a little to the east of the target, or to the right, from your perspective. From a satellite above Earth, the shell you fired travels in a straight line, directly north from you. From the perspective of someone sitting on the target, the shell veered sideways and fell to the target's left. This pattern of deflection is the Coriolis effect. This effect works in all directions. For example, if you now fire a shell 35 km directly south to another target, you are rotating faster than the target and the shell

lands to the east again, but this is now to the left of the target (from your perspective) and the shell veered *to the right* from the perspective of the person sitting on the target. Winds and oceanic currents travelling from regions of higher pressure to regions of lower pressure are subject to Coriolis forces as they head towards the region of lower pressure. Consequently, they spiral, in a clockwise direction in the southern hemisphere or in an anti-clockwise direction in the northern hemisphere.

Ekman transport occurs when a wind persists for a day or more in a consistent direction, let's say north-to-south, where the wind pushes the top several metres of water north to south, but with a veer to the right because of the Coriolis effect. This water [that is moving north-to-south with a veer to the right] must be replaced. When this occurs along the east coast of the Americas, the land mass prevents water moving from the west and so cold, nutrient-rich water wells-up from the deeper ocean, thereby producing upwelling, bringing colder water to the coastline.

Turbulence in oceans is vitally important to the functioning and behaviour of oceans. The upward movement of cold, nutrient-rich water from depth to the shallow, nutrient-depleted warm upper layers influences the productivity of the oceans (in terms of, for example, fish population size), and regional weather patterns, because of the exchange of heat between deep cold layers, warm upper layers, and the atmosphere. Turbulence even affects the "global conveyor belt", a constantly moving system of deep-ocean circulation driven by differences in temperature and salinity across large oceanic distances (Fig. 1). This conveyor belt transfers vast amounts of heat from the warm tropics to the cold northern and southern oceans and causes recirculation of enormous quantities of water between polar and equatorial regions of the oceans.

In 2006, a team in Canada (Kunze et al., 2006, Science) showed that when krill ascend towards the upper layers of the ocean at dusk, the turbulence they generate is 1000 – 10,000 times larger than when they are not ascending, and that this turbulence could bring nutrients and CO_2 from deep layers to the upper layers. Vertical migration of fish, krill, squid, and jellyfish (see below) is well-documented and in terms of biomass movement, it is the largest migration event in the world. The reason for the daily ascent is probably to avoid predators and to access new food resources and is triggered by changes in sunlight and by the biological clocks of the organisms involved. Solar eclipses can induce the ascent by fooling the organisms into "thinking" it is night-time. This large-scale vertical daily migration was discovered by the US Navy during WWII when testing sonar readings of the ocean.

Two scientists (Kakani Katija and John O. Dabiri) at the California Institute of Technology in Pasadena demonstrated (2009, Nature) that the swimming of vast numbers of jellyfish contributes as much energy into the oceans in terms of mixing of water and dissolved nutrients as that contributed by tidal movements and wind. Thus it appears that biogenic (induced by living organisms) mixing is likely to be as important as tides and wind in oceanography.

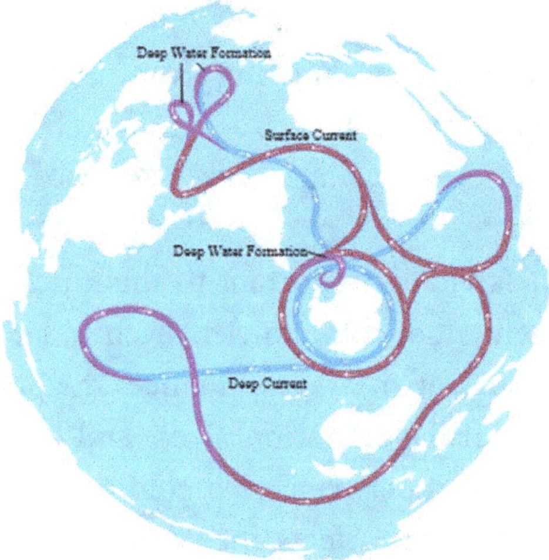

Figure 1: A map of the "global oceanic conveyor belt", drawn from a dolphin's perspective, that is, where the oceans are shown as a single body of water and the flux can be understood more easily without cutting it anywhere. This is based on a previous work of the world's oceans, by Avsa. This file was derived from: Oceans-image.svg, CC BY-SA 3.0, https://commons.wikimedia.org/w/index.php?curid=8385268 elt This image is licensed under the Creative Commons Attribution-Share Alike 3.0 Unported license.

GROUNDWATER, TIDES, AND GRAVITY

Tides have an interesting and little-known effect on groundwater located underground in aquifers. Groundwater is probably one of the least commonly understood water resources of the world (Chapter 5). About 40 % of the world's land surface is arid and receives less than 500 mm of rainfall per year and about 40 % of the world's population live within these regions. Without plentiful rainfall, human consumptive needs, including livestock watering and crop irrigation, are reliant on groundwater stores. Not only is there a large human need for water, but also a very large proportion of native vegetation is reliant on shallow groundwater stores that are tapped by roots. Strangely, the roots of trees, which can travel deeper into soil than those of grasses and shrubs, often transport water up at night from deep underground and release it into shallower layers of soil (first demonstrated by my wife in her PhD years some 40 years ago, a phenomenon called hydraulic lifting), and then they, and shrubs and grasses, reabsorb this water during the day.

Groundwater is often held between two relatively impermeable layers of rock or clay. Such *confined* aquifers (Chapter 5), when located close to the coast, respond to tidal movements, which pressurise water in the aquifer and cause water levels in wells tapped into the aquifer to rise and fall with the incoming and receding tides, showing a 6-hour cycle corresponding to the 6-hour cycle of the tides, but with a small time-lag of less than two or three hours. There isn't an increase in the volume of water in the aquifer, it is just that the weight of the incoming tide increases the pressure inside the confined aquifer, and this pushes water up higher in the wells (bores).

Not only do tides affect coastal confined aquifers, the Moon and the Sun also affect groundwater in the same way that they affect the oceans. The volume of water in aquifers can be vast. The Great Artesian Basin aquifer of Australia lies under about one quarter of Australia (about 1.7 million km^2) and contains almost 65,000 km^3 of water (Chapter 5). Such a large mass of water (it weighs about 650 thousand billion tonnes) is subject to tidal forces from the Moon and the Sun but there is an additional effect of the expansion and contraction of the Earth's crust as it too responds to the pull of the Moon and the Sun.

Tides also affect the Moon's distance from Earth and the rate of rotation of Earth. Tidal pulls of the Moon and the tidal bulge of Earth causes the Moon to be constantly moving to a slightly higher orbit and Earth to be decelerated in its rotation. What this means is that the Moon causes the tides, but the rotation of Earth pushes the tidal bulge forward slightly, because of the frictional drag of the ocean floor on the water above it. The tidal bulge now sits slightly ahead of the Moon. The tidal bulge exerts a small, but significant pull on the Moon, accelerating it by a small amount, but sufficient to make the Moon travel faster, thereby making the Moon move away from Earth, just a little (about 38 mm y^{-1}).

As the Moon moves further away from Earth, it slows down too, just a little. This is true of artificial satellites launched into space. Satellites closest to Earth move faster around Earth than satellites further from Earth. Since the Moon is about 384,000 km from Earth, this movement away from Earth is not exerting much effect on Earth. The rotation of Earth is also declining, by about 2 milliseconds (ms) per century (this rate of decline is changing over time). This means the Earth's daylength is getting longer, by about 2 ms per century. This decrease in rotation time is due to the Moon exerting a force on the tidal bulge, mentioned earlier. Given enough time (about 50 billion years) Earth and the Moon will enter tidal lock, and each will see the same face of the other for the rest of eternity – except the Sun will of course have completed its life cycle and exploded by then, so they won't.

WHY DOESN'T THE MOON FALL TO EARTH – AND WHAT WOULD HAPPEN IF IT DID?

Earth pulls on the Moon and the Moon pulls on Earth, so why doesn't the Moon fall to Earth? And what would happen if it did? Before tackling this thorny issue, I shall briefly discuss the origin of the Moon.

THE ORIGIN OF THE MOON

Despite repeated claims to the contrary, Earth has only one moon: The Moon! The Moon is about one-quarter the diameter of Earth (comparable to the width of Australia) and is the largest natural satellite in the Solar System relative to the size of its planet, or the fifth largest satellite in the Solar System overall. It orbits Earth at an average distance of 384,400 km and its gravity slightly increases the Earth's daylength. How does the Moon's gravity do this? The Moon's gravity pulls on Earth's oceans, resulting in high tides on opposite sides of the planet as the bulge in the oceans moves around Earth. Friction between oceanic water and the sea floor (which are moving relative to each other) act as a small brake, slowing the rotation of Earth by about 1.7 milliseconds each century. But back to the main question: where did the Moon come from?

The Moon was formed about 4.5 billion years ago, so it is almost as old as Earth (4.54 billion years old). It is suggested that the very young Earth was hit by a Mars-sized body (diameter of about 3,000 km) called Theia. This impact caused a vast quantity of material to be ejected from Earth, some of it derived from Earth itself, and some derived from Theia. This material coalesced under its own gravity, forming the Moon, at a distance that is just outside the Roche radius (or the Roche limit). This is the distance measured from Earth within which the gravitational field of Earth is sufficiently powerful to prevent a smaller body (the newly forming Moon, or proto-moon) from being held together by gravity. This is important because if the ejecta had coalesced at a distance less than the Roche radius (almost 9500 km for a solid object; almost 18,400 km for a liquid object), the proto-moon would have been torn apart by the Earth's gravitational pull. Outside of the Roche limit, the gravity of the proto-moon was sufficient to hold it together against the gravitational pull of Earth.

So, back to the original question: why doesn't the Moon fall to Earth?

Before answering this question, it helps to do a simple mind experiment......

Consider a rifle, perfectly mounted horizontally, 1 m above the ground. Assume the muzzle velocity is 1000 m s^{-1}. Fire the rifle. Assuming no effects from air resistance, and no objects in the way, the bullet will have travelled *about* 450 m in 0.45 s. At precisely the same time as firing the rifle, you drop a bullet from a height of 1 m above the ground. Which bullet (the one fired from the rifle or the one dropped from your hand) will hit the ground first?

First, how long before the dropped bullet hits the ground? The answer to this is determined using the following equation (sorry, but it's a simple one):

$$T = \sqrt{\frac{2d}{g}}$$

In words this equation can be written as time (T, in seconds) equals the square root of (2d divided by g), d is distance to fall (1 m in our case), and g is acceleration due to gravity on Earth (9.81 m/s/s, also written as 9.81 m s^{-2} or 9.81 m/s^2).

So, $$T = \sqrt{\frac{2 \times 1}{9.81}} = \sqrt{\frac{2}{9.81}} = \sqrt{0.2039} = 0.45 \text{ seconds}$$

Intuitively, we might expect that the bullet fired from the rifle is still travelling happily through the air after only 0.45 seconds.

In fact, they hit the ground at the same time (ignoring air resistance and the curvature of Earth – more on that important point later). Why is this? The short answer is, "gravity exerts no influence on the horizontal movement of the bullet, it only influences the vertical movement (that is, the fall) of the bullet".

The longer answer is: gravity is the same for both bullets. The acceleration due to gravity, directed downwards, is 9.81 m s^{-2} for both bullets. The equation above applies to both bullets, and both bullets must drop the same amount (1 m) before hitting the ground. So, both bullets take 0.45 s to fall to the ground. The bullet fired from the rifle is travelling at 1000 m s^{-1} as it exits the barrel of the rifle, but gravity isn't bothered by this fact. It still pulls the bullet down at the same rate (g). After 0.45 s the fired bullet has fallen 1 m and hits the ground. It probably still has forward momentum and will skip along the ground for a bit until it has lost all its energy and comes to a stop. But it did hit the ground after 0.45 s. Of course, in real life, the fired bullet is impeded by air resistance, so its velocity declines as it travels through the air.

Another quick thought experiment may also help understand why both bullets will hit the ground at the same time.

Imagine for a moment that you are a revered monk. You spend your days contemplating the universe on your mountain retreat at 10,000 feet. One day you see a Jumbo jet fly past, at your height, close to your retreat. Now because you are indeed a revered monk you have special abilities, so you can see inside the jet and observe all that goes on. As the jet comes into view you observe a steward serving drinks in cups – and one cup gets dropped. Now, what do you see? You see the cup fall to the floor of the cabin – **BUT** you also see that the cup has moved a considerable distance horizontally – in fact, the cup seems to move in an arc motion – both down AND sideways at the same time. So, what does the steward/passengers see? As far as they are concerned, the cup just falls to the steward's feet just as if they were on the ground. So, we see that in fact the cup is moving to Earth as normal with the usual acceleration **AND** moving sideways at the same time.

And what does this example have to do with the question "why doesn't the Moon fall to Earth?"

I shall attempt an answer.

But first, we must define what we mean by the word 'down'.

In the bullet dropped/bullet-fired-from-a-rifle thought experiment discussed above, there is so little difference in the direction of down for the two bullets that we don't really think about it. But down really means 'the distance moved from the initial starting point, moving towards the centre of Earth'. For the bullet dropped from a hand, the path followed by the bullet is 'down towards the centre of Earth'. It has no forward motion, so down is always in the same direction. But, for the bullet fired from the rifle, the curvature of Earth is ever-so-slightly important. For every kilometre the bullet travels, the surface of Earth 'falls away' by about 8 cm. Therefore, the bullet falls almost 104 cm (1.04 m) in 0.45 s, rather than 1 m (I assume here the bullet hits the ground after travelling horizontally about 450 m before hitting the ground). Similarly, for every 100 km I walk across the Great North American Plains (i.e., across a "flat" land surface), I walk down (measured from my starting point, and height is defined here by the distance between a tangent perpendicular to the surface of Earth at my initial starting point, and my new position, 100 km away), almost 800 cm, because of the curvature of Earth. This figure is larger than you might at first think – and it is because as you travel further from your original position, more and more of your journey is in fact directed 'downwards' relative to your starting position. In fact, when you have walked a ¼ of the globe's circumference, you are walking down quite a long way (relative to your stating point). So "down" does really need to be defined relative to something, just as distance does. Distance refers to the difference in position between two points separated in space. Down, in this scenario, is defined as relative to the tangent perpendicular to the Earth's surface. Hopefully, the following figure helps clarify this – clearly Earth and my stick figure and the 100 km walked are not to scale!

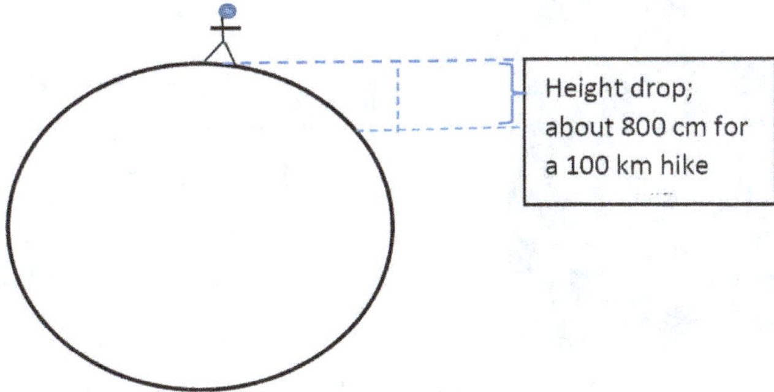

Figure 2: How far does the Earth's surface "fall away" as I walk 100 km in a "straight line" across the surface of Earth? Answer: about 800 cm. Not to scale!

Right. Back to the Moon. Imagine the Moon travelling through space, in a straight line, unimpeded by air resistance (there is essentially no air in space, which is why no one can hear you scream).

Here is the Moon, travelling in a straight line, horizontally (represented by the black arrow). This is equivalent to the bullet being fired from the rifle. Now let's introduce the sudden appearance of Earth, with its associated gravity. Clearly the images in the following figure (Fig. 3) aren't to scale in terms of diameters or distance between Earth and the Moon.

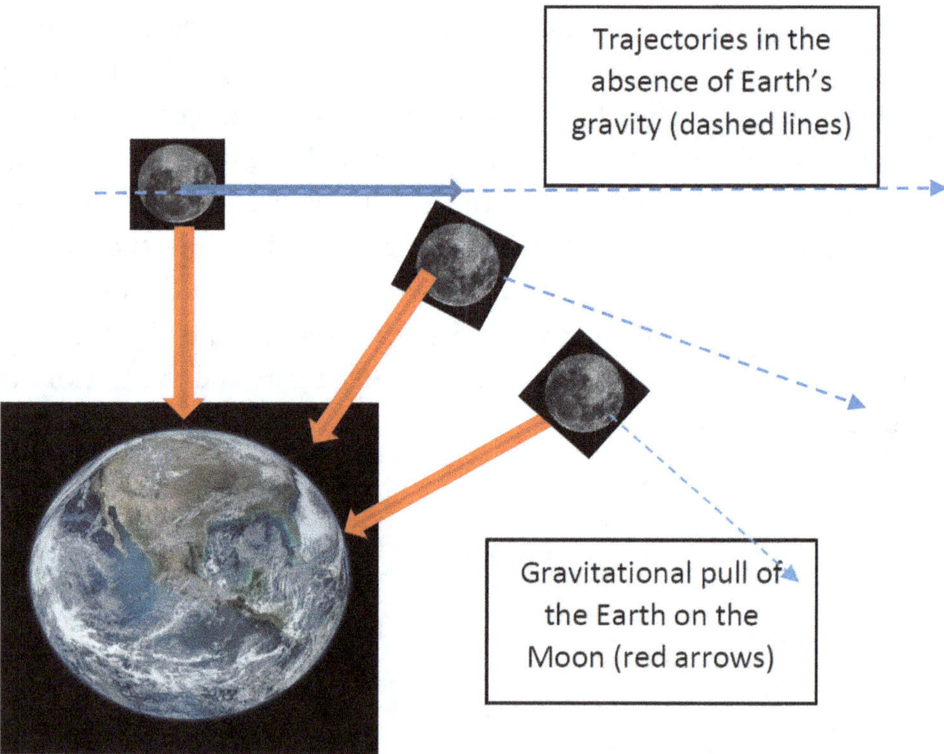

Figure 3: The Earth's gravity attracts the Moon and consequently the Moon's trajectory is altered in comparison to the trajectory it would have if Earth were absent (not to scale).

Gravity is the centripetal force that "pulls" the Moon towards Earth. A centripetal force is a force that makes a body follow a curved path. The direction of the centripetal force is always towards the centre of curvature of the path. Without this external force acting on the Moon, the Moon would keep moving in a straight line into space (we ignore the Sun's gravity here). If Earth's gravity were suddenly "switched off" the Moon would indeed hurtle out into space. Note here that we use the term *centripetal force* NOT centrifugal force – that is a whole different thing and very often misused in describing circular motions such as the Moon's orbit.

The simple analogy is spinning a mass on the end of a string. The string provides the centripetal force pulling the mass towards your hand as you spin/twirl the weight through the air. Let go of the string and the mass flies off in a straight line at a tangent perpendicular to the circle of rotation and that is because the mass is no longer constrained by the centripetal force applied by the string on the mass. This centripetal force is equivalent to the Earth's gravity on the Moon. A slingshot works on the same principle. Spin the slingshot with a stone in it, release one half of the slingshot and the stone flies off at a tangent, perpendicular to the circle of rotation, in a straight line.

For the movement of the Moon, the same issue of the Earth's curvature applies, but on a bigger scale. Remember that the Moon's orbit has an average radius of about 384,000 km, or a circumference of about 2.4 million km (i.e., the length of the orbit is about 2.4 million km). For every 1000 km travelled by the Moon around Earth (so a very small fraction of the orbit, about 0.004 %), it falls (measured relative to the start of that 1000 km journey) more than 1.3 km (Everest is about 8.8 km tall).

So, what is the answer to the question: why doesn't the Moon fall to Earth under gravity?

The answer is: it *does* fall, and it *keeps* falling. If the Moon were travelling more slowly than it currently does or was significantly closer to Earth that it currently is, it would indeed fall and hit Earth. If it were travelling much faster than it currently does, or it was significantly further from Earth (and so Earth's gravity exerted a smaller effect on the Moon) it would fly off into Space. It is the balance between Earth-to-Moon distance, gravity, and the Moon's speed, plus the curvature of Earth, that prevent the Moon from actually hitting Earth.

The Moon is almost entirely stable in an orbit around Earth. It is only "almost entirely stable" because the Moon is moving away from Earth at almost 4 cm per year. (We know this from daily laser firing at mirrors left on the Moon after the Apollo missions to the Moon). It is also important that the Moon does not interact with the Earth's atmosphere (because it is too far away) so there is no air resistance to slow it down. A bullet fired in space would travel billions of kilometres (yes, billions) because air resistance is so extremely low (almost, but not quite, zero). Consequently, it would slow down very, very, very, gradually (I am assuming it doesn't hit a planet or become affected by gravity of a planet/sun/black hole *en route*).

The Moon pulls on Earth with the same size of force that Earth pulls on the Moon (Newton's third law of motion: all forces between two objects exist in equal magnitude and opposite direction: if one object A (Earth) exerts a force on a second object B (the Moon), then B simultaneously exerts a force on A, and the two forces are equal in magnitude and opposite in direction). The Earth's gravity pulls the Moon into an approximately circular orbit and, rather surprisingly, the Moon exerts an influence on the Earth's orbit too.

WHAT WOULD HAPPEN IF THE MOON DID CRASH TO EARTH?

Let's look at two previous events that didn't end well after a celestial object and Earth had too close an encounter: the Tunguska event and the Chicxulub impact.

THE TUNGUSKA EVENT

On the morning of June 30, 1908, an estimated 80 million trees, covering more than 2000 km^2, were flattened near the Podkamennaya Tunguska River in Yeniseysk Governorate (now Krasnoyarsk Krai), Russia. This region, in the East Siberian Taiga, is, and was, sparsely populated, but some eyewitnesses were there to record the event; at least three people may have died. No impact crater was apparent then, and none has been found since, and the Tunguska blast is likely to have been an explosive disintegration of a small stony meteoroid (see below) about 50 – 200 metres in diameter, at an altitude of 5 to 10 kilometres (when it exploded). The shock wave from the air burst would have registered 5.0 on the Richter scale, and the amount of energy released is estimated to have been the equivalent of 3 – 30 million tonnes of TNT. Little Boy, the atomic bomb dropped on Hiroshima, released the equivalent of 15 thousand tonnes of TNT, or about 0.001 % of the energy released during the Tunguska blast. The Tunguska event is the largest impact event on Earth in *recorded history*, although much larger impacts have occurred in prehistoric times (see below).

Figure 4: Trees knocked over by the Tunguska blast. Photograph from the Soviet Academy of Science 1927 expedition led by Leonid Kulik. Photo taken on 30 June 1908, at 07:17, at Podkamennaya Tunguska River, Siberia, Russian Empire. From Wikipedia.

There are several hypotheses as to the exact nature of what entered the Earth's atmosphere, and what the fate of it was. In 2020, a team of Russian scientists concluded, from applying a computer model, that an iron asteroid up to 200 m in diameter, travelling at about 11 km s^{-1}, entered the upper atmosphere and then bounced out of the atmosphere, generating a shock wave that reached the Earth's surface in Siberia, knocking over the trees. At some locations, the direction of the shock wave would have been almost vertical, and trees tend to remain standing, but lose all their foliage. Further away, the shock wave would have been travelling in a more horizontal direction and trees were knocked over. This pattern in the trees was observed at the site 114 years ago. A similar pattern can be generated if the object explodes mid-air, rather than bouncing back out to space.

The question as to what the object was, remains debated. It could have been a small comet, or cometary fragment. Comets are composed of dust and frozen compounds, including ice. Comets are vapourised as they pass through the atmosphere, and therefore there would be no impact crater (but see below for a mention of this point). Glowing skies observed for a couple of nights after June 30th 1908, support the comet hypothesis because as comets break-up and their contents vapourise, the burn-up is visible at night. One possible source of this small cometary fragment is Comet Encke, which generates comet showers on an annual basis, with trajectories similar to the Tunguska event.

The chief difficulty in the asteroid hypothesis is the apparent absence of an impact crater. Perhaps the asteroid disintegrated entirely in a huge explosion, 5 – 10 km above ground, as it passed through the atmosphere, and smaller fragments burnt up, producing the bright luminescence in the night skies of Europe and Central Asia that were witnessed on subsequent nights. Or perhaps there is a crater, hiding in plain sight….

Lake Cheko is a small lake approximately 8 km north-northwest of the estimated point of the main explosion in the atmosphere. This was the site of a study undertaken by several Italian scientists.

Lake Cheko has an unusual, funnel-like, bottom morphology, with a maximum water depth of 50 m and a 0.16 depth-to diameter ratio. This morphology differs from that of subarctic Siberian thermokarst lakes (lakes formed from the melting of permafrost), which have very steep sides and flat bottoms. This morphology is compatible with the impact of a cosmic body (an asteroid), having a mass of approximately 1.5×10^6 kg (10 m diameter). The projectile that formed Lake Cheko may have been a fragment of the main body that exploded in the atmosphere. Seismic reflection profiles 10 m below the bottom at the centre of the lake indicate a sharp density/velocity contrast, consistent with either the presence of a fragment of the body, or of material compacted by the impact.

So, the take home message at this stage is that a relatively small object (50 – 200 m in diameter) was able to flatten an estimated 80 million trees across an area of more than 2000 km^2, releasing the equivalent of about 3 – 30 million tonnes of TNT. It is only the fact that Siberia was so sparsely populated that the event did not cause more loss of life and damage to infrastructure. Remember that New York covers about 800 km^2. London covers about 1500 km^2.

Now let's look at a really big impact, the meteor that wiped out the dinosaurs.

THE CHICXULUB IMPACT

The Chicxulub impactor was approximately 11 – 81 km in diameter. When it hit Earth, about 66 million years ago (Mya), it released the equivalent of 21–921 billion Hiroshima atomicbombs (between 1.3×10^{24} and 5.8×10^{25} joules; equivalent to about 1 x 10^{16} tonnes of TNT). The impact produced a core of superheated plasma exceeding a temperature of 10,000 °C and created a crater with a diameter of about 100 – 150 km wide and 20 – 30 km deep. Several rings of ejected material were formed around the crater. Under a thin ring of overlying material is a layer of melt rock exceeding 120 m in thickness. This was deposited the day after the impact. The pressure pulse and winds would have removed soils and destroyed vegetation and animals close to the impact, possibly up to 1500 km away. Much of the debris ejected by the impact fell-to-earth within several minutes, in the Gulf of Mexico and the Caribbean. The depth of the ejecta that landed quickly exceeded several hundred metres. Even 500 km from the impact site, deposits of 50 – 300 m have been detected. Currently, the crater is mostly covered by sea and overlain by 600 m of sediment.

Where is this crater? The Chicxulub crater is buried underneath the Yucatán Peninsula in Mexico. The centre of the crater lies offshore, near the communities of Chicxulub Puerto and Chicxulub Pueblo (hence the name of the crater). The date of the impact coincides with the Cretaceous–Paleogene boundary (or the "K–Pg boundary"), a little more than 66 Mya. This impact is credited with causing an intense and prolonged global-scale disruption to climate, which then caused the Cretaceous–Paleogene mass extinction event, whereby about 75 % of all plant and animal species became extinct, including all non-avian dinosaurs. How might this have happened?

The impact created mega-tsunamis. The heights of these are debated, with estimates ranging from 100 m to 1.5 km. Multiple tsunamis were able to reach the coasts of all continents, with two main mega-tsunamis generated. The first arose from the direct impact of the incoming asteroid or comet and the associated formation of the crater, and the second arose from the outward expansion of oceanic water after filling the crater. Vast landslides at coastal sites created secondary tsunamis and multiple tsunamis presumably traversed back-and-forth across the oceans for some considerable time.

The impact also created winds of speeds exceeding 1000 km h^{-1}.

Perhaps more catastrophic than the tsunamis were the massive clouds of hot dust, ash, and steam released from the crater site. Remember that the impactor carried on downwards many kilometres into Earth, within a second or two. Ejected material derived from Earth and pieces of the impactor were ejected out of the atmosphere, but re-entered Earth and become red-hot, and were distributed around Earth. Shock waves throughout the upper crust triggered global earthquakes (magnitude 12 at the impact site, magnitude 9 globally) and subsequent volcanic eruptions, including the Deccan Traps flood basalt eruption (see below), which may have occurred around the same time. Fossil evidence for a near instantaneous die-off of a range of animal species was discovered in a 10-cm thick layer of soil in New Jersey, 5,000 km from the impact site. The narrowness of the layer tells us that death and subsequent burial under debris was rapid and occurred quickly over long distances.

The ejection of millions of tonnes of dust and ash into the Earth's atmosphere would have circulated around the globe for many years. While ash falls to the ground within several days, the vast amounts of SO_2 ejected would have been converted to sulphate aerosols, which can remain in the atmosphere for years. These aerosols reflect a significant fraction of incoming solar radiation, which both cools the Earth's surface and reduces the rate of photosynthesis of vegetation. The cooling and reduced productivity of vegetation resulted in the loss of herbivores, and subsequently, carnivores, including all non-avian dinosaurs, but also, it is estimated, 75 % of all species globally, went extinct. This is one of the five great mass extinction events recorded over the past 450 million years. These events are:

1. The Ordovician-Silurian extinction (450 – 440 Mya); 60 – 70 % of all species went extinct.
2. The late Devonian extinction (375 – 360 Mya); 70 % of species lost.
3. The Permian-Triassic extinction (252 Mya); 70 % of terrestrial species, 96 % of marine species, lost.
4. The Triassic-Jurassic extinction (213 Mya); 75 % of all species went extinct.
5. The Cretaceous-Paleogene (or K–Pg extinction) event (66 Mya); 75 % of all species extinct.

Some scientists suggest a 6th mass extinction event – the Anthropocene Extinction – is caused by humans (since 1900) and remains ongoing.

Several volcanic eruptions throughout the past century have similarly caused global air temperatures to decrease. One of the largest, and best documented, was the eruption of Mount Pinatubo on June 15, 1991. This was one of the largest eruptions of the twentieth century and injected 20×10^6 tonnes of SO_2 into the stratosphere at an altitude exceeding 14 km. This eruption cooled the Earth's surface for three years following the eruption, by as much as 0.7 °C.

CHICXULUB AND THE 5TH MAJOR MASS EXTINCTION

Did the Chicxulbub cause the 5th major extinction event? The timing looks reasonable.

The age of the rocks marked by the impact is approximately 66 Mya, the end of the Cretaceous period, and the start of the Paleogene period. This coincides with the K–Pg boundary, the geological boundary between the Cretaceous and Paleogene and the K–Pg mass extinction event. Many reviews have concluded that this impact caused the 5th mass extinction.

But an alternative hypothesis to explain the 5th mass extinction exists, the so-called Deccan Traps (Chapter 6) volcanism hypothesis......

DECCAN TRAPS VOLCANISM

The Deccan Traps is in west-central India (17–24°N, 73–74°E) and is a vast accumulation of igneous rocks, called a large igneous province (LIP). LIPs are formed when magma travelled through the crust and erupted at the Earth's surface as a mantle plume moved up from depth, or where two continental plates are moving apart. The Deccan Traps (DTs) in India are one of the largest volcanic features on Earth, consisting of several layers of solidified flood basalt. These layers exceed 2,000 m in depth and currently cover an area of about 500,000 km^2. The DTs may once have covered about 1,500,000 km^2. More than 90 % of the lavas of this province probably erupted across a period of around one million years, by plume volcanism, and the timing of these eruptions coincided (approximately) with the Cretaceous-Paleogene (K–Pg) boundary (about 66 Mya). Could this volcanism have had a major role in the end-Cretaceous mass extinction?

The massive volcanic activity that generated the DTs would have released vast volumes of SO_2, ash, and dust. Unlike the Chicxulub impact, which was a one-off event, the volcanism of the DTs continued for up to 1 million years. The long-term impact on global climate must have been extreme and long-lasting. However, the volcanism associated with the DTs started perhaps 250,000-to-1 million years prior to the K–Pg boundary, and hence many years prior to the mass extinction event. It is difficult, therefore, to state that the volcanism associated with the DTs caused the 5th extinction event (see below).

Recent work published in the Geological Society of America Bulletin, in 2015, suggest that three critical events occurred across a short period of time (100,000 y). These events were the 5th mass extinction (the K–Pg boundary extinction), the Chicxulub impact and a period of especially voluminous, but brief, pulse of eruptions towards the end of the period of the formation of the Deccan continental flood basalt province. Importantly, Mark Richards and the team that published the work highlight the fact that Deccan volcanism started well before the mass extinction event and of itself, did not cause the mass extinction. Similarly, the impact of the Chicxulub on global climate was too short-lived to have caused the mass extinction alone. However, the Chicxulub impact may have triggered the short (in geological timescales) period of high-volume eruptions towards the end of the period of time when the Deccan continental flood basalt was being formed, resulting in a series of lava flows that account for >70 % of the total DTs lava flows. It was this that may have caused the 5th mass extinction. The cause of the increased volcanism following the Chicxulub impact was the propagation of earthquakes from the impact crater. There is no doubt that the energy involved in the Chicxulub was sufficient to trigger earthquakes globally and there is no doubt that earthquakes trigger volcanic eruptions.

The beginning of the extinction of non-avian dinosaurs may have started about 66.4 Mya and hence may correspond to the surge of DT volcanic activity (66.3~66.413 Mya). An increasing number of species of dinosaurs died out (along with many other types of organisms, including all tetrapods weighing more than 25 kilograms, all plesiosaurs and mosasaurs, most teleost fish, sharks, molluscs (especially ammonites, which became extinct), and many species of plankton) with the continuation of volcanic activity, and no non-avian dinosaurs survived beyond the K–Pg boundary. Non-avian dinosaurs became extinct, DT volcanism continued, and the emergence of large mammals occurred much later, but DT volcanic activity may not only have caused the extinction of non-avian dinosaurs, it may also have delayed the radiation of mammals. Radiation here means the expansion, or increase, through time, of the number of species present.

BACK TO THE MOON

The Chicxulub impactor was an object less than 100 km in diameter. The Tunguska event involved an object much less than 1 km in diameter. The Chicxulub impact triggered tsunamis, earthquakes, massive volcanism, and started the K–Pg mass extinction event 66 Mya. The Tunguska event wiped out 80 million trees, covering more than 2000 km^2.

The Moon is about 3,474.2 km in diameter, or about 34 times the diameter of the Chicxulub impactor. I think we should be immensely grateful that the Moon is not ever going to impact the surface of Earth. If it did……. we would not survive! Neither would much (anything?) else.

WHY DO WE ONLY EVER SEE ONE SIDE OF THE MOON, IF IT IS SPINNING?

Oh yes, a quick aside: the Moon is spinning on its axis, but we only ever see one side of the Moon (the near side, as opposed to the far side; by the way, the far side of the Moon is not dark), no matter when we look at it. How can that be?

The moon orbits Earth every 27.3 days (the lunar month). However, the Moon also spins once on its axis at the same rate as it rotates around Earth (that is, one spin every time the Moon orbits Earth). This means that the same face of the Moon is always facing us. I find this hard to visualise in my head, and if you do too, look at this webpage where there is a moving graphic showing this phenomenon: https://svs.gsfc.nasa.gov/4709#:~:

This phenomenon is a special case of gravitational locking (also called tidal locking) and results in synchronous rotation (where the same side of the Moon faces Earth as it rotates around Earth). The Moon used to spin much faster, but Earth's gravity slowed it down until it was in synchronous rotation.

To summarise, this chapter has discussed the following:

Tides are the result of tidal forces acting on our seas and oceans. Most of this gravity (about 54 %) originates from the Moon, but the remaining 46 % originates from the Sun. When the Sun, Earth and the Moon are aligned, tides are much larger than when they are not aligned.

Tides are important in mixing deeper and shallower layers of water, but it turns out jellyfish are as important as tides in this mixing process. Strange but true.

Groundwater (Chapter 5), that huge body of water sitting in aquifers beneath the Earth's surface, is also subject to tidal forces. Changes in the volume of water in aquifers can be detected using satellites.

The Moon is falling towards Earth – but do not panic. Fortunately, the Moon is travelling at such a speed, and is at such a distance from Earth, that it can't hit Earth. If the Moon was travelling much faster than its current speed, it would escape Earth's gravity and fly off into Space. If it travelled much slower, or was much closer to Earth, it would indeed hit Earth. If the Moon did impact Earth, we can safely assume that life on Earth would cease and Earth would change its orbit, probably into an unstable orbit, eventually spinning away from the Sun.

Previous impacts of large chunks of rock include the Tunguska Event in Siberia which released energy equivalent to 3 – 30 million tonnes of TNT and destroyed a huge area of forest. The Chicxulub impact in the Gulf of Mexico released energy that was the equivalent of 21 – 921 billion Hiroshima atomic bombs and killed the dinosaurs, because the amount ash and dust ejected high into the atmosphere was sufficient to block a large fraction of sunlight from reaching the Earth's surface, thereby starving the dinosaurs and reducing surface temperatures for several years.

The Moon is rotating but we only ever see one side of the Moon. The reason for this is that the Moon is gravitationally locked – the Moon orbits Earth every 27.3 days, but the Moon rotates on its axis at about the same rate, so we see the same side of the Moon every night. The Moon used to spin much faster, but Earth's gravity has caused it to slow down, and it is now in synchronous rotation.

THE END

WHY DOES ICE FLOAT AND HOW HAS THIS AFFECTED THE EVOLUTION OF LIFE ON EARTH?

Ice (frozen water) floats. We all know that. The North Pole (the Arctic) is an island of ice floating on seawater. Ice in my vodka floats. However, most liquids, when frozen, become more dense and sink when dropped into the liquid form. But not water. Why not? Perhaps more importantly, what would happen if ice didn't float? Turns out that there would be very profound impacts on life. It also true that Earth has indeed frozen almost entirely three times in the past 2.4 billion years (when Earth was known to be a slushy ice ball or a snowball). Why did Earth freeze over three times in the past 2.4 billion years, and are we really currently living through an ice age? These are some of the key questions to be examined in this chapter.

DENSITY, TEMPERATURE, AND FLOATING ICE

The simple answer to the question "why does ice float?" is that ice is less dense than water and therefore floats in water. Less dense means that there is less mass per unit volume, hence density can have units of g cm^{-3} or kg m^{-3} (also written as g/cm^3 or kg/m^3). These changes in density are discussed in the next several paragraphs. But why is ice less dense than water? And how does it matter for life on Earth?

The density of water at 4 °C is 1 g cm^{-3} (or 1 kg per litre or 1 tonne per m^3). As temperature increases above 4 °C, the density of water declines a little (the density of water at 20 °C is 0.9982 g cm^{-3}). This is a common response of many liquids to an increase in temperature. For example, the density of ethanol (the alcohol in the beer, wine, and spirits we drink) decreases from 0.8029 g cm^{-3} at 4 °C to 0.7852 g cm^{-3} at 25 °C. The density of solid (frozen) ethanol is 1.025 g cm^{-3} and consequently a frozen cube of pure ethanol will sink in liquid ethanol because the frozen cube is more dense than the liquid.

So why is the density of ice (0.9167 g cm^{-3} at 0 °C) less than the density of liquid water (1 g cm^{-3} at 4 °C)?

At room temperature, let's say 20 °C, water molecules are whizzing around at almost 2000 km h^{-1}. They don't get far before they hit another water molecule and indeed each molecule might experience more than 2000 billion (2 x 10^{12}) collisions each second. As temperature declines to 4 °C, the speed of the water molecules declines significantly, and the distance travelled between collisions also declines. This means the average distance between adjacent water molecules is reduced with a drop in temperature, so the number of water molecules packed into each cm^3 of water increases. Another way of saying this is that as temperature declines from 20 °C to to 4 °C, the volume occupied by 1 g of water also declines. This is illustrated in Figure 1a. Since density is mass divided by volume, the density of water increases with decreasing temperature, but only down to 4 °C. Simply put, there are more water molecules (and hence a larger mass of water) packed into each cm^3 of space.

As temperature declines between 4 °C and 0 °C, something strange happens. The density of water decreases, rather than increases, due to an expansion in volume of the water molecules, that is, for the same given mass of water, the volume this mass occupies increases (which is why water pipes burst when the water in them freezes). This expansion in volume with declining temperature is extremely unusual in liquids and explains the decrease in density we observe in water in this temperature range. The mass of water in a given volume is less at 1.5 °C than it is at 3.5 °C, for example. If you look closely at the curve for the region 4 °C to 0.1 °C in Figure 1a, you can see the small increase in specific volume (the volume of a specified mass of water) that occurs here.

In Figure 1b, we see a dramatic (about 9 %) increase in specific volume when water transitions from being a liquid (let's say, at positive 0.1 °C) to a solid (that is, ice, let's say, at minus 0.1 °C). This large increase in specific volume means that the density of ice has declined significantly and consequently ice floats in water.

This does raise a ticklish question about 4 °C. Why does the change in behaviour occur at this temperature? I shall return to this question shortly.....

The velocity of the water molecules is still decreasing as the temperature drops from 4 °C to 0 °C but other forces come into play. To understand what forces, we need to know a little about the molecular structure of water.

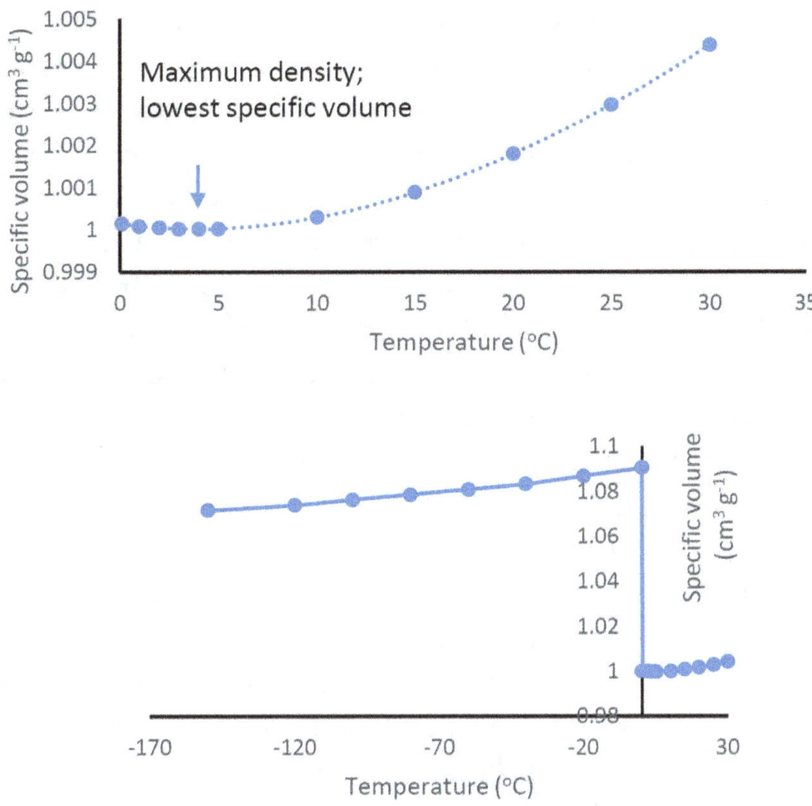

Figure 1: (a) Changes in the specific volume of liquid water as a function of temperature. Specific volume declines with decreasing temperature but only to 4 °C. Further reductions in temperature to 0.1 °C, for example, result in a slight increase in specific volume and hence decline in density; (b) a large discontinuity occurs when water freezes (at 0 °C) and there is a large increase in volume and hence a large reduction in density, so ice is less dense than water and therefore floats in water.

Water is composed of two hydrogen atoms bonded to a single oxygen atom:

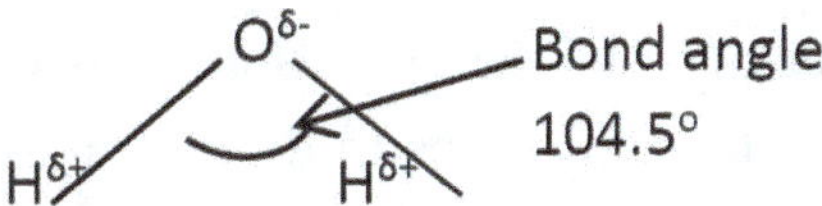

Figure 2: A crude representation of a water molecule, highlighting the bond angle and the small positive and small negative charges associated with hydrogen and oxygen atoms, respectively.

More importantly, the hydrogen atoms (H) possess a small positive charge (δ^+) and the oxygen atoms (O) have a small negative charge (δ^-) (because the oxygen atom is able to attract a larger share than the hydrogen atoms of the negative charge of the electron cloud of water molecules). This makes water a polar molecule, which is why it is so good at dissolving compounds like salts and sugars. This is an extremely important feature for life on Earth because all biochemistry happens with water as the solvent. Opposite charges attract each other in an analogous way to how N and S poles of magnets attract each other. The two hydrogen atoms of a water molecule can each attract an oxygen atom of two different water molecules, while the small negative charge on the oxygen atom can attract one of the hydrogen atoms of a molecule of water. The attraction between O and H of *different* water molecules is through H-bonds. H-bonds are the mutual attraction of a positive and negative charge on H and O atoms of different water molecules. Covalent bonds (between H and O within a single molecule) are very strong because they involve sharing of electrons between the two atoms being bonded. In contrast, H-bonds are weak and short-lived, being repeatedly formed and broken. In liquid water, water molecules are relatively free to move around, and the molecules have only a semi-ordered state and each molecule is H-bonded to, on average, 3.4 other water molecules at any one time. Molecular-scale packing (the spatial arrangement of molecules) is relatively efficient in liquid water. In contrast, in ice, H-bonding takes on a larger role in molecule-to-molecule interactions because the temperature is colder and so thermal motion is reduced, allowing the relatively weak H-bonds to exert a larger effect. At 0 °C a crystalline ice structure (I_h; see below) is formed with strong and much more stable H-bonding. This allows the water molecules to take a tetrahedral formation, with 109° angles for the four apices of the tetrahedron. Thus, on average, in ice, each water molecule is H-bonded to 4 other water molecules, but this arrangement results in a more open geometry. Consequently, the packing density of water molecules (the number of molecules per unit volume) in ice is reduced and hence the density of ice is less than that of liquid water. The closer or more open packing of water molecules in liquid and solid water, respectively, is illustrated below.

The ice that we are all familiar with (I_h) is in fact only one of 17 known solid forms of water, and is known as hexagonal ice. By experimentally varying the temperature and pressure of water it is possible to produce the other 16 ice types, including amorphous ice, rhombohedral ice, tetragonal ice, and orthorhombic ice.

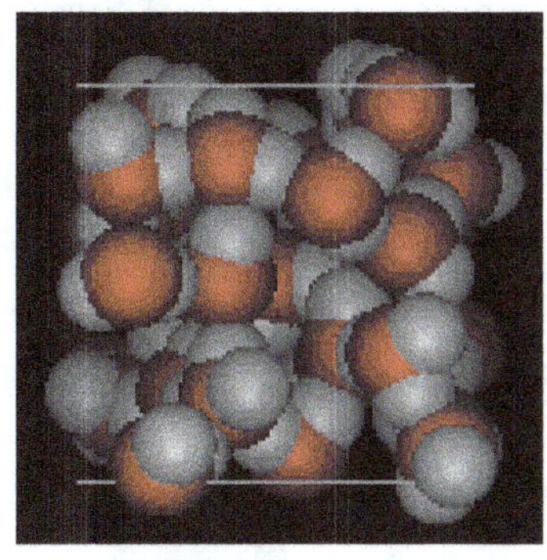
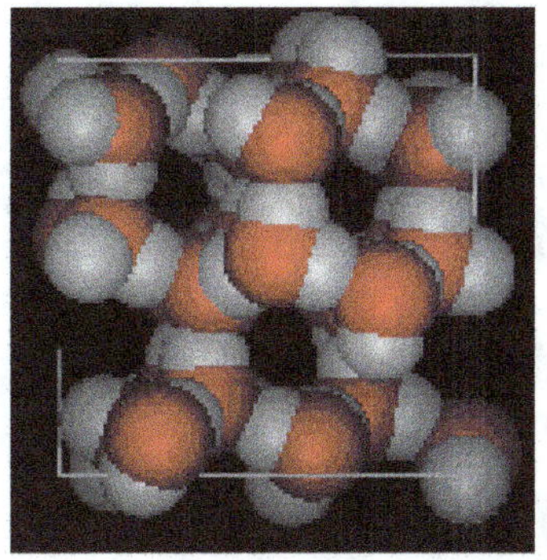

Figure 3: The density of water molecules (number of molecules per unit volume) at 4 °C is larger than the density of water molecules at 0 °C, which is why ice floats. From https://commons.wikimedia.org/wiki/File:Liquid-water-and-ice.png Created by P99am under Creative Commons Attribution-Share Alike 3.0 Unported license.

SO – WHY IS THE DENSITY OF WATER MAXIMAL AT 4 °C? PUT ANOTHER WAY, WHY IS THE SPECIFIC VOLUME OF WATER LARGER AT 3 °C AND 5 °C, COMPARED TO THAT AT 4 °C?

As temperature increases above 4 °C, the thermal energy of each molecule of water increases. Their speed increases and the distance travelled between collisions increases. The distance between individual water molecules increases and so the specific volume increases and therefore density decreases (that is, the mass of water per unit volume decreases with increasing temperature). As temperature declines below 4 °C, the opposite occurs: the thermal energy of each molecule of water declines and their speed and the distance travelled between collisions declines. But, in addition, the low energy packing arrangement of water molecules in the 0 °C to 4 °C temperature range is moving towards the more open spatial arrangement seen in ice (Fig. 2). Most importantly, the relative importance of *H*-bonds increases significantly in the range 4 °C to 0 °C and at the microscopic (that is, molecular) scale, increasingly large numbers of water molecules begin to arrange themselves into ice-like structures, held together by increasingly important *H*-bonds (the importance of *H*-bonds increases at lower temperatures because the thermal motion of molecules is less at lower temperatures so the *H*-bonds can "hold together" these ice-like arrangements for longer and longer periods of time). But why 4 °C? It is merely because the trend to move towards a larger and larger fraction of ice-like local domains (that is, large numbers of adjacent water molecules) as temperature declines below 4 °C is balanced by the opposing trend for water molecules to move around more energetically as temperature increases above 4 °C. At 4 °C these opposing trends are balanced.

WHY IS IT IMPORTANT TO US THAT ICE FLOATS?

When lakes, rivers, and sea water freeze, we see the ice floating on top of liquid water. This ice acts as a very effective insulator to the water below. The thermal conductivity of air is about 0.025 W m^{-1} C^{-1}. By comparison, the thermal conductivity of aluminium exceeds 200 W m^{-1} C^{-1} and for copper it exceeds 350 W m^{-1} C^{-1}. Metals, as we know, are good conductors of heat. Hold one end of a metal rod in a flame, and we soon feel the temperature of the rod warm up in the hand holding the other end. Concrete has a thermal conductivity of only about 1 W m^{-1} C^{-1}. It's a poor conductor of heat. But concrete doesn't float. Ice has a thermal conductivity of around 2 W m^{-1} C^{-1}, much lower than that of metal and not much worse than that of concrete. This is why the air temperature at the North Pole can be -40 °C but the temperature of the sea beneath the icecap is about -1 °C (the salt in seawater means it doesn't freeze at 0 °C). The floating surface ice insulates the water beneath it, hence the water beneath remains liquid. This is, of course, hugely beneficial to the crustacea, plankton, fish, and mammals that live under the ice. If ice sank when formed, it would be possible for more, and more, and more ice to form at the surface until all the water in Canadian, northern European and Russian, deep lakes and most, if not all, of the Arctic Ocean would become frozen, rendering life impossible for these life forms in these regions. Since a major protein source for humans for the past 1000 years in the Arctic Circle was derived from fish and whales, the absence of these would have a dramatic influence on the native peoples of the Arctic.

IMPACTS ON GLOBAL CLIMATE OF AN ENTIRELY FROZEN ARCTIC OCEAN

The impacts of a completely frozen Arctic Ocean, however, extend much further than effects within the Arctic Circle. Vast ocean currents connect the Atlantic Ocean and the Pacific Ocean to the Arctic Ocean. Cold, and hence less salty, water, enters the Arctic Ocean through the Bering Strait between Alaska and Siberia. Warmer and saltier water enters from the Atlantic between Norway and Greenland. The Beaufort Gyre (gyre means a rotating current; gyres rotate clockwise in the northern hemisphere and anti-clockwise in the southern hemisphere) "captures" this input of relatively fresh water from the Pacific but inputs of water from the Atlantic and Pacific Oceans must be balanced by losses from the Arctic Ocean via the Greenland Sea and through Baffin Bay and the Davis Strait. The complete freezing of the Arctic Ocean, arising from an "ice sinks" world, would completely change these current flows. The importance of vast ocean currents on regional and global weather cannot be overstated. For example, the Gulf Stream is a warm and rapidly moving (2.5 m s^{-1}) ocean current that starts around Florida and travels up the east coast of the US and then splits into two, one branch moving east towards Europe, the second travels south down the west coast of Africa. The climate of all Western Europe is much warmer than it otherwise would be because of the heat energy carried in this current from tropical to polar regions. Indeed, palm trees grow on the southwest coast of Scotland because of the warmth carried there by the Gulf stream. The southeast coast of Scotland, in contrast, is bereft of palm trees!

Currents in oceans can be driven by differences in density of seawater. Cold water is more dense than warm water, salty water is more dense than fresh water. When water freezes, it tends to exclude salt from the ice, so the remaining unfrozen water is saltier than before the freezing started. If the entire Arctic Ocean froze because ice sank rather than floated, water temperatures would be colder, and the surrounding water would be much saltier. Both effects influence the density of the remaining water, thereby having major impacts on the flow of deep ocean currents from polar to equatorial regions (and *vice versa*) and the transfer of energy from tropical to polar regions. In fact, it is thermohaline gradients (thermo = temperature; haline = salt concentration) that drive the biggest, deepest, and longest ocean currents around the world. Also, of course, there would be the simple physical effect of a giant frozen ocean preventing the movement of water from the Atlantic and Pacific Oceans into a frozen Arctic Ocean. Regional climates, the movement of heat energy from tropical to polar regions, nutrient movements, and global fisheries would all be strongly affected by a frozen Arctic Ocean.

SNOWBALL EARTH AND THE EVOLUTION OF LIFE ON EARTH

So, we know why ice floats and we know that it is pretty important that it does, but it turns out that a very large fraction (> 70 %) of the Earth's surface has been covered by ice, to a depth of 1 – 2 km, which equates to about 500 – 1000 million km^3 of water, at various points in geologic time. *If* the entire surface was covered, we could refer to "Snowball Earth"; if only a majority was covered (with a narrow band of open sea in a belt around the equator), we could refer to "Slush ball Earth". Whether Earth was a snowball, or a slush ball, during periods of global glaciation, remains debated, but there is no doubt that Earth was indeed one of these (most probably a slush ball, not a snowball) on several occasions and this dramatically affected the evolution of life on Earth. In this section I look at glaciations, the causes of glaciations, and the impacts of glaciations on the evolution of life.

The Huronian glaciation (or Makganyene glaciation) extended from 2.4 billion years ago to 2.1 billion years ago. In geology, billions of years ago is written as gigayears (Gya) ago. Both billion and giga refer to 1,000,000,000. Hence we can say the Huronian glaciation extended from 2.4 Gya to 2.1 Gya, of the Paleoproterozoic era (Earth is about 4.5 billion years old, or 4.5 Gya years old, give or take a month or two). The Huronian glaciation led to mass extinctions on Earth, for reasons that are initially counter-intuitive. Before the Huronian, life on Earth was dominated by anaerobic, single-celled organisms. Methane was the main constituent of the atmosphere. Methane is a potent greenhouse gas, absorbing infrared radiation about 30 times better than CO_2. Earth was therefore a warm planet. But then, cyanobacteria evolved, and oxygenic (oxygen-producing) photosynthesis began. These organisms found a new niche – utilising sunlight to fix CO_2 – and their population density increased very rapidly. The oxygen they produced is celebrated in the naming of a geological event of profound importance: the Great Oxygenation Event, when atmospheric oxygen concentrations increased, resulting in a decreased atmospheric concentration of methane. The oxygen combined with the methane to form carbon dioxide and water, which do not retain heat energy as effectively as methane, causing the cooling of the Earth's atmosphere, leading to the Huronian glaciation. During the Great Oxygenation Event, solar radiation levels were much lower than the present, meaning the Earth's temperature was far more sensitive to changes in greenhouse gas concentrations than the present, further exaggerating the cooling of Earth and the expansion of snow and ice from the poles to the Equator. The dramatic change in climate, plus the switch from an anaerobic to an aerobic atmosphere (that is, a large increase in the concentration of oxygen in the atmosphere), resulted in mass extinctions.

This Huronian glaciation period wasn't the only time a global ice-age occurred and was not the only time such glaciations were associated with major impacts on the evolution of life on Earth. The Cryogenian (cryo = cold; genian = giving birth) is a geologic period that lasted from 720 Mya to 635 Mya. The Sturtian (720 – 660 Mya) and Marinoan (650 – 635 Mya) glaciations occurred during this period and these glaciations covered most of the Earth's surface, leading to Slushball or Snowball Earth conditions. The Sturtian and Marinoan glaciations are also associated with changes in evolution. The Snowball Earth hypothesis suggests that glaciation continued over millions of years until a volume of volcanically-derived CO_2 accumulated in the atmosphere sufficient to warm the atmosphere and hence begin the melting of snow and ice. A feedforward loop was thereby created because as the total global surface area of ice and snow declined, the amount of solar radiation reflected to space declined, so more solar radiation was absorbed by the surface of Earth, leading to more warming, and more melting, and more radiation absorbed … you get the picture. There are, however, five theories proposed to explain the coming and going of Snowball Earth. These are:

(a) greenhouse gas concentrations;
(b) true polar wander;
(c) geomagnetism;
(d) galactic control; and
(e) orbital forcing (Milankovitch cycles).

I briefly discuss each of these theories in turn.

Greenhouse gas concentrations in geological history were controlled by continental plate tectonics and plume tectonics (a plume is a giant blob of molten mantle moving out from the central core of Earth towards the surface of Earth). Erosion of the Earth's crust and the subduction (the downward movement of oceanic lithosphere of one tectonic plate under an adjacent tectonic plate) of carbonate-bearing sediments into the mantle reduces atmospheric CO_2 levels, while the transfer of mantle CO_2 into the atmosphere through magmatism (the placement of magma within and at the surface of the outer layers of Earth, which solidifies as igneous rock), and volcanoes, which release CO_2, increase atmospheric CO_2 levels. The break-up of the super-continents into the continents we are familiar with today generated massive levels of erosion and the subsequent movement of sediments into the oceans and caused a colossal increase in the length of the coast, which allowed for increased coastal erosion and deposition of sediment into the oceans. The important thing to note is that these sediments are cemented by a carbonate matrix, which consumes atmospheric CO_2, resulting in a decline in atmospheric CO_2 levels and a consequent cooling effect (Chapters 11, 12). Subsequent collisions of tectonic plates, for example of the Indian Plate and the Eurasian Plate, which generated the Himalayas, also increases the generation of sediment into oceans and a drawdown of atmospheric CO_2.

The second theory, true polar wander, is a complex mechanism that leaves more questions than answers, remains poorly supported by observational data, and I do not discuss this any further.

Changes in the magnetic field around Earth, that is, the third theory, strongly affects the number of cosmic rays passing through the Earth's atmosphere, which strongly affects the amount of cloud cover. Cosmic rays are high-energy protons and atomic nuclei that move through space at nearly the speed of light (Chapter 6). They have multiple origins, including the sun, from outside of our solar system but within our own galaxy, and from distant galaxies. Sustained declines in the magnetic field are associated with large and sustained increases in cloud cover, thereby reducing the amount of solar radiation arriving at the surface of Earth, thereby cooling Earth. There is extensive evidence that the Earth's magnetic field is not constant and can decline by 100 % (that is, decline to zero) for short periods of time before either reversing in polarity (the North and South poles flip) or increasing again (Chapter 6). In the last 3000 years, for example, the Earth's magnetic field has declined by about 30 %. The Earth's magnetic field deflects most of the solar wind, a stream of charged particles that can strip away the ozone layer that protects Earth from harmful ultraviolet radiation and which is associated with skin cancer and damage to cellular DNA (Chapter 6).

The fourth theory, called the galactic control mechanism, proposes that periods of intense gamma ray release (as occurs during the birth and death of stars) results in the production of large numbers of cosmic rays, which result in extensive and prolonged cloud cover, which cools Earth, resulting in glaciation of Earth. In addition, our solar system bobs up-and-down through the narrow Galactic Disc (the Milky Way is a spiral disc galaxy that is about one thousand light years thick) as we travel through the Universe. It takes about 60 million years to move from one side of the disc to the other and it extends about 200 light years above, and then below, the spiral disc of our galaxy (Fig. 4). It has been proposed that the cosmic ray concentration "above" the disc is much larger than "below" the disc and this is likely to have a profound impact on the Earth's climate. Interestingly, the fossil record may also have some periodicity within it, with a 60-million-year periodicity, reflecting our movement up-and-down relative to the spiral arm of our Milky Way. Currently, we are "above" the disc, and heading further out, so cosmic ray concentrations are likely to increase for the next 100 million years or so and hence we may expect to see increased cloud cover and reduced temperatures on Earth over this time-scale (going to be a long wait).

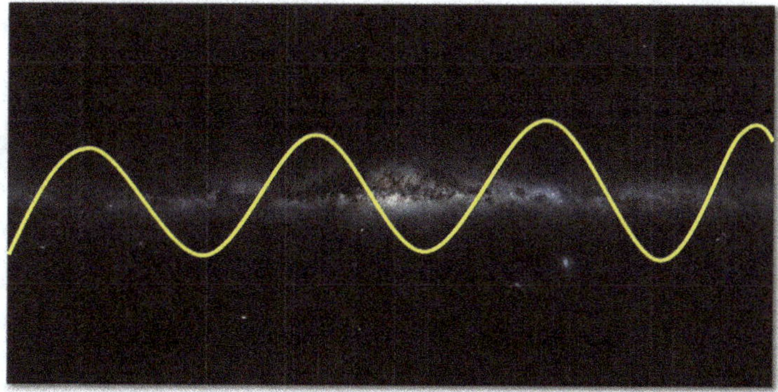

Figure 4: A 360-degree panoramic image of the Milky Way, covering the entire southern and northern celestial sphere, reveals the cosmic landscape surrounding Earth. The plane of our Milky Way Galaxy, which we see edge-on from our perspective on Earth, cuts a luminous swathe across the image. The author of this book added the crude yellow line to represent the movement of Earth above and below the Galactic Plane. The image (without the yellow line) was created by the European Space Agency: https://www.eso.org/public/images/eso0932a/
Image reproduced from https://en.wikipedia.org/wiki/Milky_Way

The fifth theory is based upon orbital forcing (the Milankovitch cycles; Fig. 5; also see Chapter 11). The Milankovitch cycles describe the effects of slow changes in the movement and orientation of Earth on Earth's climate over thousands of years. Earth's orbit varies between almost circular to more elliptical (its *eccentricity* changes). When the orbit is more elliptical, seasonal variation in the distance between Earth and the Sun is larger than when the orbit is more circular. Consequently, the amount of solar radiation received at the surface of Earth varies more between seasons when the orbit is more elliptical. On an annual time-scale, changes in eccentricity are tiny and seasons don't reflect tiny changes in eccentricity, but over tens-of-thousands of years, the accumulation of small changes in eccentricity mean that seasonal variation in distance between Sun and Earth become more significant and affect climate more strongly. The rotational tilt of Earth (its *obliquity*) also changes slightly. A larger tilt makes the seasons more extreme. Additionally, the direction pointed to by Earth's axis changes (*axial precession*), while Earth's elliptical orbit around the Sun rotates (*apsidal precession*). The combined effect is that proximity to the Sun occurs during different astronomical seasons. Changes in the solar radiation received by the Earth's surface because of these long-period cycles cause gradual, but significant, changes in the average temperature of the Earth's surface. However, it is important to note that Milankovitch cycles cannot explain the current (20th, 21st century) changes in climate because the changes in global climate are occurring too quickly to be a result of any of the three cycles, or their interactions. Later, I discuss the Milankovitch cycles in relation to the *current ice age* that we are living in.

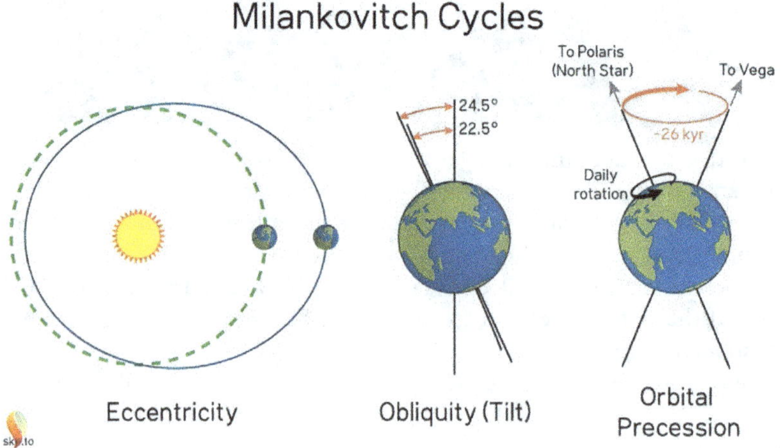

Figure 5: A representation of the Milankovitch cycles present in the Earth's movement around the Sun. From https://skepticalscience.com/graphics.php?g=342. Created under Creative Commons Attribution v4.

These three cycles (eccentricity, obliquity, precession) have different periodicities. The major component of eccentricity has a periodicity of 413,000 years but minor components have 95,000-year and 125,000-year cycles (with a return period of 400,000 years). These major and minor cycles combine into **a 100,000-year cycle**. The angle of Earth's axial tilt (the obliquity) exhibits a cycle of about 41,000 years. Axial precession, the trend in the direction of Earth's axis of rotation has a periodicity of 25,771.5 years. Most important is the observation that ice age cycles of the Quaternary glaciation over the last million years have a **periodicity of 100,000 years**, which matches the cycle of eccentricity.

SNOWBALL EARTH AND THE CAMBRIAN EXPLOSION

Life prior to the Proterozoic was primitive and unicellular. Nucleic acids (RNA, DNA; Chapter 7) were not contained in a membrane-bound nucleus and intracellular organelles (for example, chloroplasts, mitochondria) were absent (this is the definition of prokaryotes – the absence of intracellular membrane-bound organelles and nucleus). Bacteria are prokaryotes.

The **Proterozoic**
The Proterozoic is a geologic period running from 2,500 Mya (2.5 Gya) to 538.8 Mya. It is subdivided into three geologic eras (from oldest to youngest): the Paleoproterozoic, Mesoproterozoic, and Neoproterozoic.

The Proterozoic covers the time **from the appearance of oxygen in Earth's atmosphere to just before the proliferation of complex life** (The Cambrian Explosion – see Textbox below).

Several events have been identified as occurring during this eon, including the transition to an oxygenated atmosphere during the Paleoproterozoic; the evolution of eukaryotes; several glaciations during the Cryogenian Period in the late Neoproterozoic Era; and the Ediacaran Period (635 to 538.8 Mya), when soft-bodied multicellular organisms first evolved.

Eukaryotes are organisms having a membrane bound nucleus containing nucleic acids, and membrane bound sub-cellular organelles, including chloroplasts, mitochondria, and endoplasmic reticula. They can be unicellular and small, or multicellular, and often very large. Humans are eukaryotes, as are trees, elephants, and fungi.

Eukaryotes emerged about 2.1 Gya to about 1.8 Gya, that is, shortly after the start of the Proterozoic. It has been proposed that the end of the Huronian glaciation (or Makganyene glaciation; the first slushball Earth), that extended from 2.4 Gya to 2.1 Gya, in the Paleoproterozoic era, may have been a trigger for the emergence of eukaryotes. Similarly, the Neoproterozoic Snowball Earth (Sturtian (*ca* 720 – 660 Mya) and Marinoan (*ca* 650 – 635 Mya)) Snowball Earth periods have been proposed as the trigger for the Cambrian explosion (see following text box).

What might explain this sudden explosion in animal diversity? Several hypotheses have been proposed….

The first hypothesis highlights the massive increase in atmospheric oxygen concentration that began around 700 Mya, providing a vital compound required for cellular aerobic respiration. High rates of aerobic respiration enable rapid movement and the evolution of more complex body structures. A correlation between atmospheric O_2 concentration and biodiversity of eukaryotes may have existed early in their evolution. Increased atmospheric concentrations of O_2 also favours the evolution of larger (and hence more complex) animals because respiratory demand for O_2 increases with animal volume faster than the rate at which surface area increases (volume is a third power of radius but surface area is only a square power of radius). This means that a larger atmospheric O_2 concentration is required to overcome the diffusional limitation to the movement of O_2 into the centre of a larger animal.

The second hypothesis proposed that the massive extinction of life that occurred because of the Sturtian (*ca* 720 – 660 Mya) and Marinoan (*ca* 650 – 635 Mya) Snowball Earth periods, (just prior to the Cambrian), created a vast number of unoccupied niches (ecological spaces that a species can occupy) into which new species could evolve. Empty niches are extremely attractive because of the absence of competitors for food and other resources. Furthermore, during periods of Snowball Earth, massive areas of rock erosion caused by glaciers may have increased the nutrient content of seas, which began to appear across the globe as ice melted globally. Shallow seas were initially formed, with warm and light-filled water, thereby supporting the evolution of animals feeding off algae. Volcanically active rifts may also have released significant quantities of calcium and other nutrients that are required to build hard exoskeletons of early marine life and eventually the internal skeletons of vertebrates.

The Cambrian Explosion, or the Biological Big Bang

The Cambrian Explosion, or the Biological Big Bang, occurred about 538.8 Mya to about 513 Mya (these timings are debated), at the beginning of the Cambrian Period. The Cambrian Explosion was characterised by a massive and rapid (in geological terms) explosion in the number of animal phyla found in the fossil record. An extremely good example of the Cambrian explosion is the Burgess Shale in British Columbia, which was deposited in the middle-Cambrian, when the "explosion" was already several million years old. This formation contains the first appearance in the fossil record of brachiopods (marine invertebrates having a hard upper and lower shell (as opposed to a hard left and right shell, as in bivalves)), with clam-like shells, as well as trilobites, molluscs, echinoderms (marine invertebrates that include starfish, brittle stars, sea cucumbers, and sea urchins), and many odd animals no longer with us, including Opabinid, with five eyes and a nose like a fire hose. Perhaps most importantly, during the Cambrian Explosion, all the key elements of all the major formats of the "floorplan" of animals we see today were established. Bilateral symmetry (a left and right side), the development of "heads" and tails" and appendages, also appear. The first vertebrates appeared about 530 Mya.

Figure 6:   Image by Nobu Tamura (http://spinops.blogspot.com) - Own work, CC BY 3.0, https://commons.wikimedia.org/w/index.php?curid=19462324

The last hypothesis to explain the sudden explosion in animal diversity proposes that changes in the genetics of pre-Cambrian animals may have included the slow evolution of a "genetic tool kit" of genes that govern physiological and morphological development. Small changes to small families of genes can have very large impacts on development, allowing natural selection to select the successful changes. This hypothesis suggests that this was a period of extensive evolutionary experimentation, and once the most successful body-plans and life-histories had evolved, further experimentation had far less impact on speciation and biodiversity. An increase in cosmic radiation levels at Earth's surface has been documented for the period 900 Mya – 600 Mya, which may have contributed to a high rate of mutation and hence accelerated the potential for increased biodiversity. Of course, the evolution of a bilateral body plan, internal skeletons, and the entire class of vertebrates (which arose about 450 Mya) is rather central to the later development of hominids (about 15 – 20 Mya) and modern humans (about 0.8 – 0.3 Mya).

THE MOST RECENT ICE-AGE: 2.58 MILLION YEARS AGO, TO THE PRESENT

The Quaternary Period began 2.58 Mya and continues to the present. The start of the Quaternary is defined by the formation of the Arctic ice cap. Throughout the Quaternary Period, glaciers and ice sheets sitting on land (as opposed to floating on an ocean) have been present – think Greenland and Antarctica for extensive terrestrial ice sheets, and glaciers in New Zealand, the Andes of S. America, and Europe. Consequently, geologists conclude that we are currently living through an ice-age, although throughout the Quaternary, the areal extent of ice sheets and glaciers has waxed and waned. During glacial periods, temperature declines and ice sheets and glaciers expand. The initial trigger for the Quaternary glaciation was the long-term decline in the concentration of CO_2 in the atmosphere due to weathering of the Himalayas. The weathering of rock has long been known to affect atmospheric CO_2 concentrations through the Urey reaction, whereby $CaSiO_3$ reacts with two CO_2 molecules from the atmosphere and water to yield (via carbonic acid) $CaCO_3$ (limestone), SiO_2, one CO_2 and one H_2O molecule. Consequently, one CO_2 molecule is removed from the atmosphere for every molecule of $CaSiO_3$ that reacts. The $CaCO_3$ flows into rivers and eventually the oceans, where some is used to construct hard shells of marine organisms and some sinks to the bottom of the ocean where it forms limestone sedimentary rock, trapping the CO_2 for millions of years. The weathering of rock (not just the Himalayas) has been proposed as a major cause of large changes in atmospheric CO_2 concentrations for the past 56 million or more years. Research is underway to examine the possibility of spreading finely-ground calcium and magnesium-rich silicate rocks on agricultural land to remove vast quantities of CO_2 from the atmosphere via the Urey reaction as a means of tackling the rapid rise in atmospheric CO_2 levels that have occurred over the past 250 years.

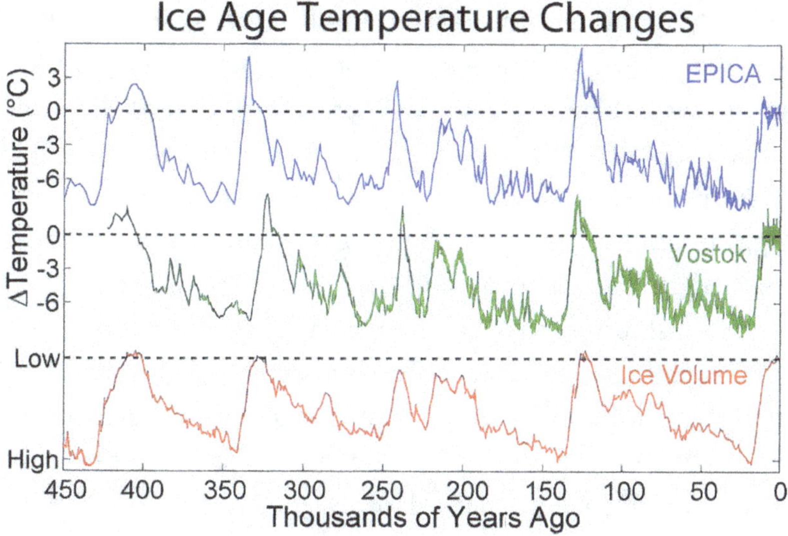

Figure 7: Antarctic temperature changes during the last several glacial/interglacial cycles of the present ice age and changes in global ice volume. The present day is on the right of the x-axis. The blue and green curves show local changes in temperature at two sites in Antarctica. The red curve shows a reconstruction of global ice volume and is scaled to match the scale of fluctuations in Antarctic temperature. Note that changes in global ice volume and changes in Antarctic temperature are highly correlated. Horizontal dashed lines indicate modern temperatures and ice volume. Image from https://commons. wikimedia.org/wiki/File:Ice_Age_Temperature.pngCreated under the GNU Free Documentation License, Version 1.2.

Over the past 1 million years, the advance and retreat of glaciers and ice sheets has been dominated by a 100,000-year cycle (Fig. 7), with 40,000-year and 21,000-year cycles superimposed upon the longer cycle. The first 2 million years of the Quaternary, however, were dominated by the 40,000- and 21,000-year cycles. We are currently in an interglacial period and the last glacial period ended about 11,700 years ago (the current geological epoch is called the Holocene and started 11,700 years ago; a glacial period is a period of time (lasting thousands of years) within an ice age characterised by colder temperatures and the advance of glaciers).

The causes of the advance and retreat of glaciers and ice sheets during the quaternary include many of those causing the start-and-end of periods of Snowball Earth, and can be summarised thus:

1. The Milankovitch cycles (see the discussion earlier in this chapter), that set of cyclic variations in the characteristics of Earth's orbit around the Sun, are strongly implicated in determining the timing of glacial and interglacial periods within a single ice age (for example, the current ice age we are living through). The influence of these cycles is termed "orbital forcing" and this is a widely accepted explanation (when coupled to feedback and feedforward processes).

2. Changes in atmospheric greenhouse gas composition, especially CO_2 and CH_4, arising from volcanism and the weathering of rock.

3. A feedforward loop, especially increases in Earth's albedo (surface reflectance) arising from increases in the areal extent of snow and ice, is required to amplify an initially small change in snow and ice cover, which result in an ever-increasing area of ice and snow because more-and-more solar radiation is reflected to space. However, feedback loops, which dampen the change arising from an input signal, also exist at the same time. An example of a feedback loop is the increase in global aridity that occurs as the areal extent of ice sheets and glaciers increase, because less water evaporates from ice and snow than from liquid water (oceans) and bare (wet) soil, and because the lower temperature associated with increased ice cover result in reduced rates of evaporation. Consequently, the deposition of water onto ice sheets and glaciers declines and gradually these go into retreat. Feedback and feedforward loops dominate at different times, leading to the advance and retreat of glaciers and ice sheets across geological time.

In conclusion, this chapter has demonstrated the very unusual property of water when it freezes, namely, the large expansion in volume upon freezing (water is the only non-metallic chemical to do this). This explains the reduction in density (mass divided by volume) of ice compared to water and explains why ice floats. It is extremely fortuitous that ice does float because it acts as a thermal insulator to the water below, allowing life to survive the periods of global glaciation – snowball or slush ball. These periods of global cold have been associated with dramatic impacts on the evolution of life, including the Cambrian Explosion (or the Biological Big Bang). During the Cambrian Explosion, all the important elements of all the major "floor-plans" or body-plans, of animals living today were established, including bilateral symmetry, the development of "heads" and tails" and appendages. The first vertebrates appeared about 530 Mya, for which we should be very grateful. Curiously, we appear to be living through an ice-age.

THE END

CHAPTER 4

We all know water freezes at 0 °C, right? This was certainly central to the discussion of Chapter three. However, in that discussion, I was talking about our common experience of the freshwater we find in our taps, in rivers, lakes and non-saline groundwater. However, when we discuss *absolutely* pure water, perhaps 0 °C is not the correct answer. Everyday measurements of everyday water that demonstrate a 0 °C freezing temperature for water suffer from the presence of two important artefacts, which I discuss later. So, what is the freezing point of pure water when these two artefacts are removed?

Water does undoubtedly freeze, and many species of plants and animals must survive extremely cold winters without dying, despite containing a lot of water. How do they stay alive? What is their secret? Indeed, how do some animals survive being frozen solid and yet 'come-back-to-life' after thawing? Can this teach us anything about cryonics, the process of freezing and preserving tissues (especially human ones) and bringing them 'back-to-life?' What an eclectic mix of themes tackled in this chapter.

WHAT IS THE FREEZING TEMPERATURE OF WATER?

Everyone knows water freezes at 0 °C (-32 °F), at normal atmospheric pressure. Don't we? Turns out that it isn't quite as simple as that. In clouds, liquid water can be observed at -40 °C. In the laboratory you can chill *ultra-pure* water to -40 °C without it turning to ice. Our everyday experience of a 0 °C freezing temperature is, in fact, an artefact arising from impurities within the water and scratches on the inner surface of the beaker/ice tray. Dust, bacterial cells, even dissolved gases such as oxygen, and minute scratches on the inner-surface of containers, act as triggers for the formation of regions of more ordered (less randomly aligned) collections of molecules of water. It is the formation of these ordered regions that allows *H*-bonds (Figure 1 below, and Chapter 3) among adjacent molecules of water to persist for longer than usual and a localised crystalline structure appears. These localised areas produce a "run-away" effect of rapidly growing ice crystallisation (Chapter 3). Accurately determining the temperature at which this freezing first occurs (called ice nucleation) is more difficult than might be imagined because once initiated, ice formation cascades extremely quickly through the small volume of water required in the experiment (it is very hard to achieve homogeneity of temperature within a large volume of water). Also, when ice forms, it releases a small, but significant, amount of energy, which raises the local temperature of the water by a small amount, thereby confounding the experimental control of temperature.

To overcome this problem, Emily Moore and Valeria Molinero (University of Utah) used a molecular-scale mathematical model of water that replicates the all-important hydrogen-bonded structure of water and accounts for the tetrahedral configuration seen in most ice on Earth (see below). As the temperature of the water represented in the model approaches -48 °C, water molecules form tetrahedrons, with each molecule loosely bonding to four other molecules via *H*-bonds (Fig. 1). The density of the water decreases, its heat capacity increases and its compressibility increases; these changes in structure and properties control the rate at which ice forms. The results of the model simulations suggest that the rate of crystallization of water reaches a maximum around -48 °C, below which ice nuclei form faster than liquid water can equilibrate, and run-away ice formation occurs.

There is another twist to the simple question: what is the freezing point of water? When tiny (10 – 100 μm radius) droplets of ultra-pure water are chilled to between -40.15 to -38.15 °C, spontaneous freezing occurs and this has generally been *assumed* to represent the *spontaneous homogenous freezing temperature of water*. The words "spontaneous homogenous" are there to refer to the absence of impurities and surfaces acting as nucleation sites.

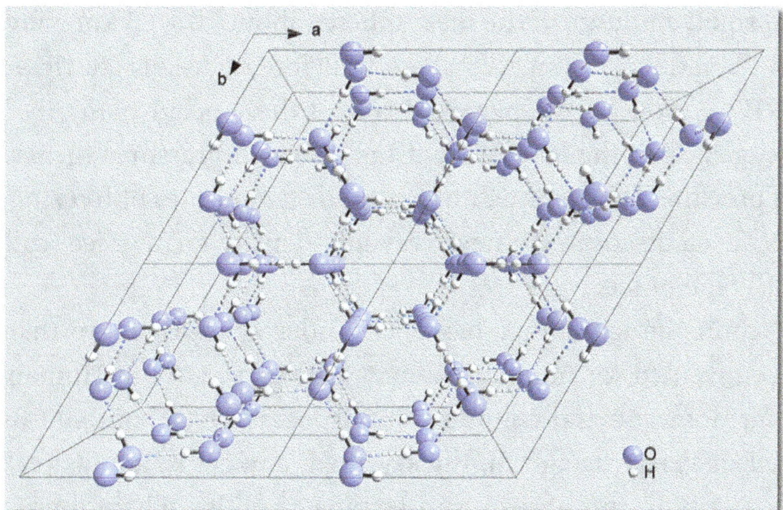

Figure 1: Crystal structure of I_h ice showing the tetrahedral structure. Hydrogen bonds are represented by dashed lines, linking the *H* (white spheres) of one molecule to the *O* (blue spheres) of an adjacent molecule. I_h ice is one type of ice found on Earth (see below). From: https://en.wikipedia.org/wiki/Ice_Ih#/media/File:Cryst_struct_ice.png

Well, even that assumption turns out to be probably wrong: it is likely that the presence of the surface tension of the microdroplets acts as a nucleation surface. The larger the drop of water in the experiment, the higher the freezing temperature, reflecting the increase in surface tension with increasing drop size. This might account for why the experimentally determined freezing temperature was a little warmer than the temperature deduced from the model simulation: the model did not include the effect of surface tension of tiny droplets of water. So, determining the actual freezing point of absolutely pure water, minus the effects of surface tension and other surfaces, remains difficult, but we know it probably lies between -38 °C and -48 °C.

HOW MANY TYPES OF WATER ICE ARE THERE?

Rather amazingly, there are 20, or perhaps 21, different types of ice in the solar system. Almost all ice on Earth is hexagonal crystalline ice, called Ice 1_h. A very small amount of ice on Earth (including the atmosphere) occurs as Ice 1_c, a metastable cubic crystalline ice, which has oxygen atoms arranged in a diamond pattern. This type of ice is produced at very cold temperatures (-143.15 °C to -53.15 °C). It may be present in small amounts in the mesosphere (about 70 – 85 km above sea-level). We then have a series types of ice called, wait for it…Ice II, Ice III, Ice IV, Ice V, Ice VI, Ice VII ….. you get the picture. None of these occur naturally. These types of ice can be created in the laboratory using different pressures (generally, but not always, high pressures, low temperatures, and different rates of freezing). A team in Japan recently even discovered a new, very low-density form of ice (called aeroice); this is the 21st type of ice.

Making ice in the laboratory has been undertaken for more than 150 years. Freezers were invented for home use about 110 years ago. But Humans have been making ice for domestic and commercial purposes since about 400 BC, when ice was produced in Persia using tall, thick-walled, conical buildings (called a Yakh-chāl; Fig. 2) sitting on top of deep sub-terranean vaults, where freezing temperatures were generated overnight through evaporative cooling. Evaporative cooling occurs when heat from the surrounding air, or an underlying surface, for example, our skin, is used to convert liquid water into water vapour; this is how sweat cools us and how plants cool their leaves during the day (via transpiration).

Figure 2: A Yakhchāl of Abarkuh, Iran. Image created by Pastaitaken. From Wikipedia; https://commons.wikimedia.org/wiki/File:Yakhchal_of_Yazd_province.jpg. This file is licensed under the Creative Commons Attribution-Share Alike 3.0 Unported license.

POLAR BEARS AND BOREAL FORESTS

It does not take long for inadequately clothed humans to die when exposed to a temperature of -15 °C and hypothermia can set in within 10 – 20 minutes. It takes a very thick fur coat, layers of fat and a metabolism that generates internal heat, for polar bears (and penguins and walruses) to cope with freezing conditions. Coniferous trees, have none of those adaptations and experience temperatures much lower than -15 °C and survive. Boreal forests (meaning Northern forests) occupy a circumpolar belt (Fig. 3) in the northern hemisphere. These are mostly dominated by evergreen needle-leaved species, including pines (*Pinus*), spruce (*Picea*), and fir (*Abies*), but also an 'unusual' deciduous conifer genus, larch (*Larix*). Boreal forests cover about 17 million square kilometres, or almost 12 % of Earth's land surface, and dominate land cover in Russia, Canada, and northern Europe. In Russia, Boreal forests are called Taiga.

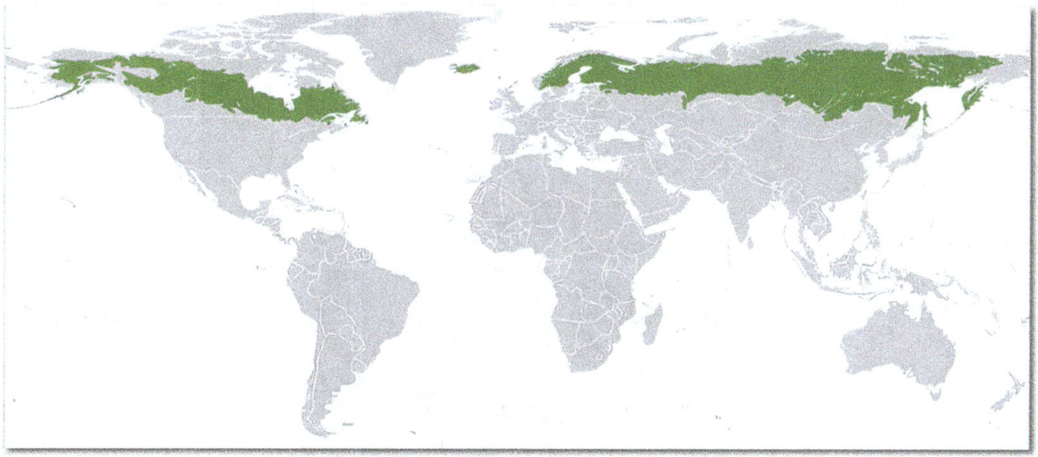

Figure 3: The global distribution of Boreal forests.

Annual average daily temperatures can be -30 °C, or lower in regions occupied by Boreal forests. In the summer, daylengths are extended, to 18 – 19 h in some locations, but conversely, during the winter, daylength can be as short as 5 – 6 h. The growing season is confined to late spring, summer, and early autumn (50 – 100 frost free days per year). Somewhat surprisingly, Boreal forests store approximately double the amount of carbon stored in rainforests. This carbon was once in the atmosphere as CO_2 but has been fixed by trees and then accumulated as wood and as soil carbon because of the slow rate of breakdown of organic material (dead foliage, branches, cones) in these locations (because of the cold). This begs the obvious question: how do long-lived trees survive the long, extremely cold, dark winters every year?

COPING WITH FREEZING WINTERS

There are three major stresses that evergreen trees in Boreal forests need to tolerate. The first is freezing temperatures (Fig. 4), and this induces two additional stresses: water stress and, during daylight hours, photo-inhibition (light-induced inhibition of photosynthesis). We won't be talking about photo-inhibition any further, except to say the needles of conifers have several (biochemical) strategies for overcoming or minimising photo-inhibition.

Figure 4: Almost all species of conifer maintain a green canopy throughout winter and are therefore frequently covered in snow. Picture by the author.

When air and soil temperatures are sub-zero for prolonged periods, the probability of the formation of ice *within* cells is high. When water freezes it expands (Chapter 3) and ruptures living cells of needles, roots, and stems. Frozen water even ruptures the dead, water conducting pipes in the stem, known as xylem. Although dead, these xylem cells are vital to the functioning of all living trees. When the ice melts, the ruptured and formerly living cells are dead, and the xylem in stems (and needles) is dysfunctional and cannot transport water. Therefore, *intra-cellular* freezing must be avoided in Boreal trees as it is inevitably fatal for the cell. As autumnal temperatures decline and day lengths shorten, evergreen trees undergo a process of frost-hardening, whereby the temperature that induces freezing damage progressively declines (that is, frost hardiness, or resistance to frost-induced damage, increases). Both signals, temperatures falling and daylength becoming shorter, are important triggers for the start of the development of frost hardiness prior to winter. Trees adapted to survive freezing in high latitudes have evolved resistance to naturally occurring temperatures as low as -70°C. Fully cold-hardened trees can exhibit freezing resistance to liquid nitrogen (-196 °C) and even liquid helium (-269 °C). To develop such resistance to extreme freezing, woody plants must be able to regulate ice nucleation, supercooling and cytoplasmic vitrification. Ice nucleation (the formation of crystals around a nucleation site, such as a bacterial cell) can be regulated by antifreeze proteins. Supercooling is the occurrence of liquid water at temperatures significantly lower than 0 °C). Cytoplasmic vitrification, or the formation of intracellular "glass" without crystallisation, occurs in the temperature range of -20 to -30 °C.

Several processes are involved in inducing frost hardiness in coniferous species. Cryoprotective processes include the following:

1. Changes in membrane composition, to allow membranes to remain fluid at low temperatures. This is analogous to the difference between margarine and butter – butter hardens at a higher temperature than margarine because of the difference in the number of double bonds between adjacent C atoms in butter and margarine. Margarine has more double bonds than butter and this determines its melting point.

2. Accumulation of solutes (for example, soluble carbohydrates such as sucrose) within cells, which depresses the freezing point of intra-cellular water by several degrees (Fig. 5). However, this is insufficient protection for anything other than the very early stages of autumnal cooling. The presence of solutes in water reduces the freezing point by hindering the development of the strong stable *H*-bonds between adjacent water molecules required for the formation of ice.

3. Most important for the development of frost hardiness are the biochemical changes that prevent intra-cellular ice nucleation. Ice nucleation is the initiation of freezing of liquid water and hence is the conversion of liquid-to-solid water. As discussed above, small particles, including bacterial cells, or dust, in otherwise "pure" water can initiate freezing very quickly. By

preventing ice nucleation, cells can tolerate temperatures as low as -38 °C, very close to the spontaneous nucleation temperature of water. An additional mechanism pushes the temperature tolerance even lower, which I discuss towards the end of this paragraph. Ice nucleation can be regulated to some degree by the action of antifreeze proteins (also called ice-structuring proteins). Antifreeze proteins are found in cells of animals, plants, fungi, and bacteria. They act by binding to very small ice crystals, stopping the run-away effect of ice crystal growth throughout the cell, which would rupture, and hence kill, the cell. In addition to the action of antifreeze proteins, living plant cells can accumulate solutes in their cell walls (that is, accumulating solutes outside of the cytoplasm of the cell). These solutes attract water from the cytoplasm of the cell, across the plasmalemma (the membrane that surrounds each living cell in plants; it presses up against the cell wall. The cell wall is therefore defined as being outside of the cell's interior, which is defined as everything found within the plasmalemma) and into the cell wall and associated air spaces. Only a very thin, (molecular-scale), layer of water, plus associated cryo-protecting solutes, is held tightly to the surface of proteins and membranes inside the cell, and this layer does not freeze. In contrast, water in the cell walls does freeze but induces no intra-cellular damage (because it is "outside" the cell). The biochemical machinery within the cell is thereby protected from freezing and these tissues can tolerate temperatures below -100 °C. Antifreeze proteins are used in the production of some ice-creams and the protein is mass produced using genetically modified bacteria.

Freezing temperatures induce water stress (an insufficiency of water) in tree canopies because intracellular water is moved out of the cytoplasm (the inside of the cell) to the cell walls. Not only that, frozen stems, roots, and the frozen upper soil profile prevent water from moving (as a liquid) up to the canopy and this intensifies the degree of water stress. However, the presence of the thin layer of *liquid* water and cryo-protectants around proteins and membranes inside of cells of needles of conifers prevents fatal dehydration.

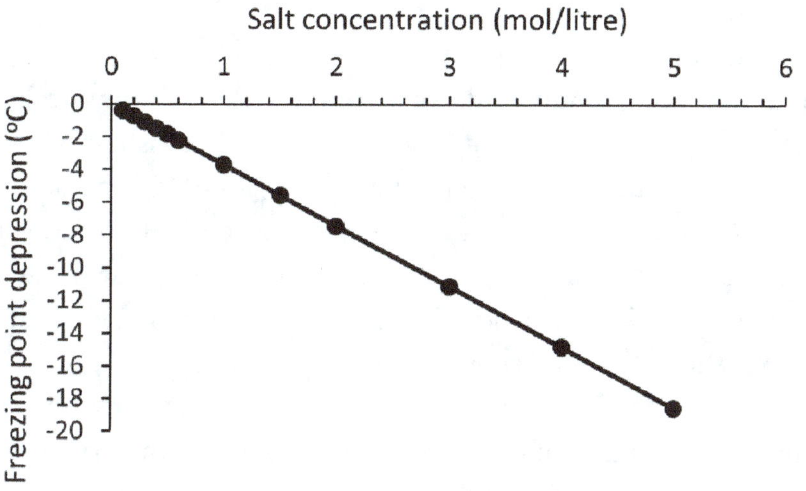

Figure 5: As the concentration of salt in water increases, the freezing point of the solution decreases. The same phenomenon occurs if salt is replaced with sugars or any other solute.

Water in the xylem (the network of water-conducting pipes linking roots to leaves/needles via the stem; xylem is dead when functioning as water-conducting pipes), if frozen, will expand and rupture xylem cells, preventing them from functioning as the pathway for water transport throughout the plant. Within the stem, xylem is found as a 1 – 6 cm wide band called sapwood (because it is part of the woody stem and it transports sap). Sapwood of Boreal trees, despite being dead, must be protected from freezing to the same degree as living cells in leaves and needles. At sub-zero temperatures water in xylem exhibits two metastable features (see textbox). First, it is supercooled (and hence its temperature is below the ice nucleation temperature). Second, it is under tension (the pull upwards on the water column, plus the pull of gravity downwards on the water column, puts the water under tension, like a stretched elastic band). These two metastable features can be maintained over time only if nucleation sites (such as gas bubbles) are absent. The temperature at which ice nucleation occurs in xylem increases as the radius of the water conducting cells of the xylem increases. Consequently, the narrowest xylem cells are found in conifers (as opposed to deciduous trees, for example, which have much wider xylem cells), conferring maximal frost resistance and thereby allowing their dominance in Boreal forests.

> **Metastable states**
>
> Metastable states of a compound, for example water, is a physical state (for example, as a liquid or solid) that, under "normal circumstances" should not be present. Thus, water in the xylem of a conifer at -10 °C would, under normal circumstances, be frozen and be present as a solid. However, this super-cooled water can stay liquid, but not for a long time, and it may freeze abruptly. Such states are called metastable. They are not equilibrium states; at negative temperatures the only *equilibrium* state of water is ice.

In an industrial, conservation or commercial setting, the cryopreservation of cells, especially plant genetic resources for future use, aims to ensure long-term storage of viable plant cells at ultra-low temperatures. The cryopreservation of plant genetic resources aims to ensure the long-term storage of viable and genetically stable plant material at an ultra-low temperature. At these temperatures, plant tissues are preserved in a state where cellular divisions and metabolic activities are minimised.

LESSONS FROM BOREAL FORESTS FOR CRYOPRESERVATION OF HUMANS

Cryonics is not the same as cryogenics, although these two words are often confused in media stories. The latter is the study of the impact of very low temperatures (lower than -120 °C) on materials (metal, gases, chemical compounds). The former is the practice/ technique of deep-freezing biota, especially the bodies of people who have just died, in the hope that future technologies and scientific knowledge may allow these bodies to be thawed, successfully, in the future, bringing the individual back to a meaningful life.

A few moderately famous people have either died and had their bodies frozen or have booked a spot in a cryonic chamber (or cryo chamber; incorrectly called cryogenic chamber in many media stories) for when they die. The corpses of two baseball players (Ted Williams and son John Williams) have been frozen, as has the body of Hal Finney, a software engineer who developed several early console games and was a strong supporter of Bitcoin. People who have booked their spot in advance include PayPal founders Luke Nosek and Peter Thiel. Jeffrey Epstein, the disgraced financier, wanted to have his head and penis frozen after death so that his DNA would live on. A very scary thought!

So, what is involved in cryopreservation? Let's start with the easiest (in terms of size and structural complexity) biological system to freeze: eggs, sperm, and embryos. As we have learned for the cells of trees of Boreal forests, the uncontrolled and extensive formation of ice within cells is invariably terminal for the cell. The expansion of water when it freezes ruptures cells and tears cell membranes, leading to cell death. Avoiding ice formation is therefore imperative. Successfully freezing eggs, sperm, and embryos has been undertaken for many decades. Embryos are frozen at the blastocyst stage and consist of about 100+ cells and are about 0.1 mm in diameter. Human spermatozoa are about 50 µm in length, while human eggs are about 100 µm in diameter (0.1 mm; the largest cells in the human body). Because these cells are really very small (compared to a human corpse), permeable cryoprotectants ("antifreeze") can diffuse into the cells rapidly and exert their protective function. Permeable cryoprotectants include ethylene glycol (a major component of antifreeze used in car radiators), glycerol (used in the food technology industry) and dimethyl sulphoxide (used in the medical industry for pain relief). Non-permeable cryoprotectants include sucrose (the principal sugar we use to sweeten our tea and coffee) and starch. These remain outside of the cell during freezing. Their mode of action is to cause water to diffuse out of the cells of the embryo prior to freezing, thereby reducing the probability of cell rupture during cryopreservation.

The original method of freezing eggs, sperm, and embryos involves the sequential addition of cryoprotectants over a period of about 10 to 20 minutes. This is followed by a gradual cooling (about 0.3 – 2 °C per minute down to -196 °C). Cooling is performed slowly to allow the cryoprotectants to enter the cells and ensure slow dehydration of the cells without formation of intracellular ice crystals.

An alternative process uses rapid freezing and the non-permeable cryoprotectants undergo vitrification (vitrification is the rapid cooling of a liquid without the formation of ice crystals). The solution forms an amorphous glass (the word glass here means an amorphous non-crystalline solid) because of rapid cooling by direct immersion of a polyethylene straw containing the embryos into liquid nitrogen (-196 °C). Non-permeable cryoprotectants also draw water out of the cells, a key step in freezing eggs, sperm, and embryos. However, it is necessary to prevent the cells from shriveling like a raisin and this is achieved by the permeable cryoprotectants within each cell. These also minimise damage to cell membranes and proteins. Thus, the process applied to successfully freeze eggs, sperm, and embryos is extremely similar to some of the steps involved during freeze-hardening of conifers (that is, accumulating "antifreeze" within cells and also outside of cells; preventing intra-cellular ice formation, and protecting membranes and proteins from being denatured). This rapid method of freezing is much less common than the slower, more traditional method.

But what of freezing corpses, or just the head of a corpse (yes, people do pay to choose this option)? As can be inferred from the discussion of freezing eggs, sperm, and embryos, the first problem is simply the fact that the corpse (or just the head or even just the brain) is just too big and too complex for cryoprotectants to diffuse rapidly into all the cells of the corpse/head/brain. And then there is the problem of freezing an object that contains tissues with very different densities (bone *versus* brain tissue, for example) and different thermal conductivities, meaning that rates of freezing differ significantly. Revival of a frozen corpse requires the repair of damage induced by years of storage without oxygen, the toxicity of the cryoprotectants used, fractures arising in tissues during differential warming, and of course, applying a cure for the original cause of death. Finally, very few companies offering cryonic preservation remain financially viable for very long. Of all the cryonics companies that were formed prior to 1973, only one remains in business to-day. It costs money to maintain a large thermos of liquid nitrogen and pay the staff and rental costs of the facility. Eventually, the money runs out….. and the corpses must be disposed of……

There is no evidence that we can cryonically freeze corpses and revive them. Although a group of professional neurobiologists recently declared there had been a successful freezing of a rabbit's brain followed by a thaw of the brain without any apparent loss of structural integrity, all the brain's proteins were later shown to have been inadvertently chemically cross-linked, making them non-functional.

But are there any animals that can experience freezing, and full revival after thawing, in the natural world?

ANIMALS THAT SURVIVE FREEZE/THAWING

Several animal species are capable of surviving freezing and thawing, including the Arctic Woolly Bear Caterpillar (*Gynaephora groenlandica*), the Darkling Beetle (*Upis ceramboides*), the Sleeping Chironomid (*Polypedilum vanderplanki;* see below) and the Wood Frog (*Rana sylvatica;* see below). Larvae of the Goldenrod Gall fly, (*Eurosta solidaginis)* are exposed to temperatures lower than -35 °C as they overwinter inside the stem of goldenrod plants (*Solidago* spp.), while larvae of *E. solidaginis* can survive exposure to -80°C. Here, I focus on the Sleeping Chironomid and the Wood Frog.

THE SLEEPING CHIRONOMID

Chironomids are small, non-biting midges, that superficially resemble mosquitos. Their larvae can be found in almost any freshwater body globally. One species, *Polypedilum vanderplanki* (common name the Sleeping Chironomid), is common in semi-arid regions of Africa. The larvae of this insect make mud tubes at the bottom of shallow pools and inhabit these tubes. However, these shallow pools often completely dry out for significant periods. Consequently, the body of the larvae may lose up to 97 % of its water. Remarkably, when the pool receives rain, the larvae rehydrate and appear to be unharmed, and metamorphosise into adult midges. The equivalent loss for a 100 kg human would be to lose about 58 kg of water (because humans are about 60 % water). Humans are extremely dead well before this percentage water loss (about 15 – 25 % water loss results in death in humans). The larvae of the Sleeping Chironomid exhibit one of the most extreme forms of anhydrobiosis of any animal. Anhydrobiosis means 'life without water' and animals that exhibit anhydrobiosis enter an extreme state of suspended animation in which their metabolism comes, reversibly, to a standstill. It has been experimentally shown that an individual larva can be dehydrated and rehydrated 10 times and still undergo metamorphosis and become an adult. Most organisms that can dehydrate to extreme levels and then rehydrate and survive are smaller than 1 mm in diameter. The larvae of the Sleeping Chironomid, while long (about 5 – 8 mm), are narrow. They are the largest metazoan (multi-celled animals with cells differentiated into tissues and organs) animals to exhibit anhydrobiosis. A slow rate of desiccation is required for the larva to survive (about 2 days to dehydrate). During this period, several genes are upregulated (become more active), and the larva accumulate a lot (20 % of their dry weight) of trehalose, a sugar made of two molecules of glucose. Trehalose is commonly accumulated by organisms to increase tolerance to desiccation. They also accumulate LEA-like proteins. LEAs are Late Embryogenesis Abundant proteins. These proteins protect against the aggregation of proteins that occurs during the dehydration induced by very low temperatures. And here comes the link to freezing tolerance.

When larvae of the Sleeping Chironomid are dehydrated to less than 8 % of their fresh (hydrated) weight, they can be immersed in liquid helium (-269 °C) for 5 minutes, thawed, and they "wake up" and carry on eating and will turn into adult midges at a later date. In its native habitat, the Sleeping Chironomid (adult or larva) do not experience freezing conditions and therefore it is hard to imagine that this tolerance of extreme cold has evolved through natural selection. Rather, this tolerance is an accidental by-product of the evolution of anhydrobiosis, which does confer a large benefit to the larvae in their natural habitat. Indeed, the Sleeping Chironomid exhibits very high tolerance to a wide range of environmental stressors, including ionizing radiation and very high temperatures (through up-regulation of heat-shock proteins; see text box). When the larva is only dehydrated to 25 % of its fresh weight, it does not exhibit extreme tolerance to freezing and does not accumulate as much trehalose as when it is dehydrated to 3 % of its original fresh weight, demonstrating the importance of the accumulation of trehalose in the development of tolerance to very low temperatures.

Heat shock proteins

Heat shock proteins (HSPs) were originally identified as a response to high temperature stress (hence their name) but they are now known to increase cellular resistance of plant cells to multiple environmental stresses, including high temperature, UV, low temperature, and salinity, and inflammation and infection in humans. HSPs often act as chaperones, guiding the refolding of proteins damaged by heat stress. They have been found in all species examined, from bacteria to fungi to plants to humans and therefore they represent a very old lineage of proteins with an important function in stress tolerance.

HSPs are useful as immunologic adjuvants, which means they make vaccines work better when they are included in the vaccine. The reason (mechanism) behind the oft-observed development of cross tolerance (whereby tolerance to water stress, for example, which can be increased by exposure of a plant to a mild water stress, *also increases* that plant's tolerance to a second stress, for example, low temperature) is most likely because HSPs are the cellular mechanism underpinning tolerance to a wide range of stressors.

Sleeping Chironomid larvae can exhibit anhydrobiosis because of several mechanisms, including the replacement of structural water with compatible solutes, including disaccharides (e.g. trehalose); the formation of stable glass-like structures from highly hydrophilic proteins which prevent the biomolecules from irreversible aggregation; upregulation of reactive oxygen scavenging compounds to reduce the damage that reactive oxygen inflicts on DNA, RNA and membranes; and the accumulation of heat shock proteins.

THE WOOD FROG

The Wood Frog (*Rana sylvatica* or *Lithobates sylvaticus*) is the only cold-blooded tetrapod living north of the Arctic Circle in the Western hemisphere. During winter it becomes dormant in leaf litter or in a shallow soil burrow. Either location lies well within the depth of soil that experience freezing temperatures. Being cold-blooded, its body temperature, especially when dormant, falls as environmental temperatures fall. To what extent does it freeze, and how does it tolerate freezing?

Extracellular freezing in the body of the frog occurs at around -2 °C. Experimentally, all frogs kept at -4 °C for 2 days recovered, and half of all those kept at -6 °C for 11 days survived (Storey and Storey, 1984, Journal of Comparative Physiology B). About 1/3rd of each of the frog's body water was frozen (at both temperatures), including gut contents, with ice crystals forming *around* the heart, liver, and gut, although these organs were not themselves frozen. The heart stops beating, and breathing stops, when the frog is frozen. Thawing over a period of about 24 hours seems to be needed for full recovery from being frozen. When extracellular fluids

Figure 6: *Lithobates sylvaticus* (or *Rana sylvatica*; the Wood Frog) found in southern Quebec. From: https://en.wikipedia.org/wiki/Wood_frog#/media/File:Ranasylvatica.JPG Image created under Creative Commons Attribution 3 unported. Author W-van.

freeze, water is drawn through osmosis from the inside of the cells, meaning that the concentration of solutes (for example, salts, sugars, amino acids) within the cells increases very significantly, and the volume of the cell declines. Both of these processes are harmful to cells and need to be minimised. The Wood Frog appears to rely almost entirely on the accumulation of glucose (although some urea is also accumulated) as cryoprotectants in blood, and all organs examined (including liver, heart, and leg muscles) and this glucose was derived from glycogen in the liver.

During thawing, the heart begins beating again, lenses in the eyes thaw (ouch), and pulmonary respiration and blood circulation begins, despite parts of the body still being frozen (one or two hours into a thaw). Movement of legs takes up to 12 h and full mobility requires up to 24 h.

Glucose has three roles in the frog during winter. First, it makes the solute concentration of blood and cell water more concentrated, and this makes the freezing point of the cell water and blood decline (Fig. 5), thereby deferring the onset of freezing slightly and reducing the volume of ice produced at any given temperature. Glucose accumulated in cells also reduces the degree of cell shrinkage during freezing. Finally, glucose is a cellular fuel that can be used under anoxic (low levels of oxygen) conditions, which occur within the frog because the heart has stopped beating, and breathing has stopped.

FREEZING/THAWING OF NON-HUMAN ANIMALS AND PLANTS COMPARED TO FREEZE/THAWING HUMANS

There are three fundamental differences between the freezing and thawing of Boreal forests/animals in nature (with survival of the tree/animal experiencing the freezing/thawing) and the cryonic freezing of humans. These are (i) trees/animals have evolved over a long time to adapt to being frozen and subsequent thawing in the field. Humans haven't evolved to survive being frozen; (ii) several physiological adaptations occur *during* the days and weeks *prior* to freezing as daylengths shorten and temperatures drop. Humans do not exhibit these adaptations. Perhaps most importantly, (iii) humans are **dead** when they are frozen, these trees/animals are not.

In conclusion, I think we can reasonably say the following:

Absolutely pure water (containing no viruses, bacteria, dust, dissolved gases) that is not in contact with surfaces such as the inside of a beaker, does not freeze at 0 °C, it freezes at much lower temperatures. There are also many forms of water ice, but one type accounts for almost all ice on Earth.

It does not take long for inadequately clothed humans to die when exposed to sub-zero temperatures. Animals exhibit multiple adaptations to survive freezing winters, including very thick fur coats, layers of fat, a metabolism that generates internal heat, hibernation or temporary migration to warmer climes. Coniferous trees have none of those adaptations and yet experience and survive very very cold temperatures. This is possible because of their own adaptations to freezing, and animals and plants that tolerate being frozen exhibit similar biochemical/physiological adaptations. Freezing temperatures do cause a small-to-large fraction of their water to freeze, but resistance to extreme freezing involves multiple freeze-hardening processes, including the ability to regulate ice nucleation, supercooling and cytoplasmic vitrification, by accumulating sugars, minimising the amount of water contained within cells and accumulating antifreeze proteins. In addition, there is a small number of animal species that can be entirely frozen and yet remain alive and "come-back-to-life" after being frozen. These animals use the same mechanisms to tolerate freezing temperatures as those employed by coniferous trees. It seems extremely unlikely that humans will be able to be frozen when dead and brought back to life later. Whether the recent findings of a cytoprotective perfusate that can be used *post-mortem* to prevent cellular damage (to improve organ transplant success) following the cessation of blood flow can be extended to sub-zero temperatures, remains to be seen. Some form of cryopreservation or suspended animation will need to be engineered if humans are to realistically probe into deep space.

THE END

WHAT IS GROUNDWATER? HOW DID GROUNDWATER INFLUENCE EARLY HUMAN EVOLUTION? WHAT IS OVEREXTRACTION OF GROUNDWATER DOING TO JAKARTA AND NORTHERN CHINA?

Groundwater is the vast, mostly hidden, store of water lying underground. More than 2 billion people globally rely on groundwater for drinking and within urban areas, about 50 % of human consumptive water use comes from groundwater stores. Of the approximately 750 km^3 (750 billion litres) of water extracted from aquifers each year, about 70 % is used for agricultural purposes. But how far back in history can we trace groundwater use? Did groundwater access affect early human evolution in Africa? Has groundwater scarcity caused civil unrest and lead to armed conflict? How has access to groundwater became weaponised? And where do we find groundwater, how is it different from just plain old soil water, and does groundwater get replenished, and if so, how? Does it matter to the ecology of a region if groundwater is extracted too quickly? Does anything live and breed in aquifers? How can two satellites help us manage groundwater and why is China moving nearly 30 billion m^3 (or 30 km^3) of water *annually* from southern China to northern China? Finally, what is overextraction of groundwater doing to Jakarta and northern China? These are the key questions that I will address in this chapter.

GROUNDWATER ACCESS WAS AN IMPORTANT DRIVER OF HOMININ (INCLUDING HUMAN) EVOLUTION

Hominins (in particular, extinct human species, including *Homo erectus* and *H. habilis*) originated in Africa and migrated out of Africa several times, starting about 1.85 Mya. The climate of the original home of hominins was warm-to-hot and arid-to-semi-arid. Landscapes in arid and semi-arid climates rarely contain perennial streams and in the East African Rift System (EARS; located in north-eastern Africa; a region where 3 tectonic plates are moving apart), perennial streams are absent, although many large, deep, and very old freshwater, saline, and alkaline (or soda) lakes are present. But the EARS is the site of early hominin evolution and the location of the origin of early (and modern humans). This begs the questions: how did early hominins obtain sufficient water in a hot and dry environment, and why and how did they migrate away from the lakes of the EARS? The EARS also experiences (and has experienced for millions of years) periods of multi-year droughts and hominins were not well adapted to coping with multi-year droughts. How did we, the hominins, survive and migrate?

Hominins *versus* hominids

Defining these two words is slightly fraught because as the number of techniques applied to, and the number of studies of, the early evolution of Hominins and Hominids increase, definitions change. Here is one, relatively modern, definition:

The word Hominin refers to the group of species consisting of modern humans, extinct human species (species of the genus *Homo*) and our immediate ancestors (e.g., *Homo erectus, Australopithecus*).

This contrasts to Hominids, which is the group of species consisting of all modern and extinct Great Apes (i.e., modern humans, chimpanzees, gorillas, and orangutans, and all their immediate ancestors). A widely used *older* definition of Hominids is the same as the more recent definition (the one given above) for Hominin, giving rise to much confusion among different sources of information.

Hominins possibly arose about 7 – 9 Mya, and Hominids arose about 1 – 2 Mya. *H. habilis* lived about 2.4 – 1.5 Mya in Africa and might have been the first Hominin to make tools. *H. erectus* lived about 250,000 years to 1.6 – 2 Mya in Africa and *H. sapiens* arose in Africa about 250,000 – 300,000 years ago and then migrated out of Africa.

During the early 21st century a multidisciplinary team of anthropologists, ecologists, climate modelers, archeologists, hydrologists, and human evolutionary biologists examined ancient sites in northern Tanzania (the Olduvai Basin), within the

EARS. They concluded that during the past *ca* 6 – 7 million years (when hominin evolution was occurring in the EARS), volcanic highlands received significant rainfall, which flowed downslope from the highlands and recharged groundwater (GW) stores in the lower pyroclastic alluvial fan (clay, silt, and sand deposited from volcanic eruptions) at the bottom of the highlands. Once underground, evaporation of this water was reduced to very, very low levels. The hydraulic pressure of the water in the alluvial fan increased down the slope, so that in some places it was forced up to the land surface, producing springs, which were likely to have flowed for 100 years, to perhaps 1000 years, even during a prolonged drought. It was these springs, dotted across the landscape that sustained hominins during the recurrent droughts of the EARS. According to many, it was the need to cross these landscapes in search of water that drove many features of hominin evolution, including bipedalism, a larger brain, and the capacity to problem solve into the future. The existence and persistence of GW discharge to the land surface within the EARS (and elsewhere across East Africa) has now been extensively documented and can be viewed as an important mechanism supporting migration, evolution, and gene transfer between isolated populations of hominins (who encountered each other at these springs and presumably inter-bred). There is increasing awareness that access to GW has been a key requirement for the evolution, persistence, and migration of hominins and early humans. To-date, more than 50 sites have been found where GW supply and hominin and archaeological remains have been identified in Africa and the Middle East (including Turkey, Iran, Jordan). The concentration of butchered bones and stone tools from about 1.8 Mya, are largest at springs in Africa, providing strong supportive evidence of the importance of access to GW to hominin behavior. At one site in the Olduvai Basin, the remains of three sympatric (co-occurring in space) hominin species have been identified (*Paranthropus boisei, Homo habilis* and *H. erectus*) at a spring.

CIVIL UNREST AND GROUNDWATER RESOURCES

Water is obviously a vital resource for all households, villages, towns, regions, and countries. It comes as no surprise, then, that access to water can be, and has been, used as an instrument of oppression and a means of fighting war.

The first documented use of denial of access to water during war occurred about 2500 years BC, when two Sumerian city states, Lagash (an ancient city state located northwest of the confluence of the Euphrates and Tigris rivers in Iraq) and Umma, another ancient city in Iraq about 35 – 40 km away, fought each other over access to water and access to fertile irrigated farmland. During the conflict, the King of Lagash drained some canals and blocked access to others, thereby depriving the city state of Umma of water and hastening their defeat. About 1800 years later, the Assyrians similarly diverted canal water and, possibly for the first time in times of conflict, poisoned and destroyed wells, denying the enemy access to fresh water.

Clearly, conflict over access to water in the Middle East and northern Africa has been a feature of these regions for a very long time. Annual rainfall across the region is low (less than 200 mm y^{-1} for much of the region) and annual evaporation rates greatly exceed rainfall (hence the predominance of deserts). Consequently, water security is particularly low. The region accounts for about 6 % of the world's population but contains only about 1 % of global freshwater resources. Power imbalances (military, economic, or control of access to fresh water) among regionally co-located countries, result in conflict. Turkey enjoys significantly better control and access to water in the Euphrates and Tigris than neighbouring countries because both rivers, which bound the fertile crescent where agriculture first became widespread about 11,000 years ago, originate within Turkey. Turkey can therefore exert significant influence on Syria and Iraq, through which the two rivers flow. Similarly, Israel exerts significant influence over water access in Palestine through its control of the Jordan River. It has been argued that conflict over access to freshwater in the region and Israel's plans to move water from the Jordan River to the Negev, to support infrastructure and an increased population density, has been central to Arab-Israeli disputes, including the start of the 6-day war. Meltwater of Mount Hermon on the border between Syria and Lebanon directly feeds GW in the region, which emerges at the ground surface at three large springs, giving rise, eventually, to the Jordan River. The Jordan River is therefore very much dependent on GW (a gaining river; see below).

Many, many, examples of denial of access to water, especially GW in arid and semi-arid regions, can be found over the past 4,500 years, including, but certainly not limited to the following:

1. The 12th-century Holy Roman Emperor Frederick Barbarossa is reported to have dropped human corpses down wells in Italy in 1155 to poison the well and deny access to fresh water.

2. Saladin, the great commander of the Saracens, deprived the Crusaders access to water in the Holy Land in 1187, by filling wells with sand.

3. During WWII, in the late 1930s and early 1940s, the Japanese military infected thousands of Chinese wells in the Pacific theater, with cholera.

4. Troops loyal to Saddam Hussein destroyed or poisoned wells in northern Iraq, including an especially large one north of Halabja, as part of a campaign targeting Kurds in northern Iraq.

5. ISIS sabotaged a large number of wells by filling them with rubble, oil, or other chemicals and often removed or destroyed the associated pump, power cables, generators and transformers required to access water from these wells.

6. ISIS seized Mosul dam, the largest in Iraq, in August 2014, and held the Fallujah barrage (on the Euphrates) from February 2014 to June 2016 and the Ramadi barrage (on the Euphrates) from May 2015 to January 2016. In 2014, ISIS used the Fallujah barrage to block an Iraqi army advance with floodwaters, and subsequently blew up most of its gates.

Given all of this, it seems to me to be of value to take a deeper (pun intended) look at GW……

AN INTRODUCTION TO GROUNDWATER

When it rains on a crop field, much of the rainwater gets stored in the upper 1 m of soil and this is readily available to roots of the crop. Water enters the upper soil profile and will continue to move down through the soil profile as long as the force of gravity exceeds the capillary force of the soil particles holding onto the water. The upper soil profile wets up to field capacity, which is the volume of water held in the soil when drainage due to the downward pull of gravity has stopped. This may take a few hours for a very sandy soil, or many days for soils with more clay. However, at field capacity, air spaces remain within the soil profile and the soil is not saturated with water. Plant roots extract their water requirements from this soil water.

This is in marked contrast to water in an aquifer, that is, GW. An aquifer is saturated with water and mostly devoid of air spaces (because they are full of water). An aquifer is a layer of permeable rock, rock fractures, or unconsolidated (loose; not compacted) materials, such as pebbles or sand, that lies underground at depths that typically range from a couple of metres to thousands of metres. Two types of aquifer exist: confined and unconfined. A confined aquifer is a layer of water-saturated permeable rock, rock fractures, or unconsolidated material sitting between an upper and a lower relatively (relative to the aquifer layer) impermeable layer of rock or clay. The water in a confined aquifer does not rapidly interact/mix with water in the upper soil profile, that is, water doesn't move between the confined aquifer and the soil layers above it when it rains. The water in a confined aquifer is generally under pressure. Punch a hole in the upper impermeable layer by drilling a bore and water moves upwards under pressure and may even reach the soil surface, generating a local spring, spa, or oasis (more on this topic later). Such an aquifer is called artesian, meaning the pressure in the aquifer is sufficient to bring the water to the soil surface if a bore is drilled. The Great Artesian Basin (GAB) of Australia is one giant confined aquifer, lying beneath about 1.7 million square kilometres of land. An unconfined aquifer has a confining layer underneath it, upon which the water sits, but it lacks the upper, low permeability, layer sitting on top. The upper boundary of saturated water is generally referred to as the water table. Water can be extracted from unconfined aquifers by digging a well to a depth that is below the water table and dropping a bucket (hopefully on the end of a length of rope) down the well and letting it fill up.

HOW MUCH WATER IS CONTAINED IN AQUIFERS, AND WHERE ARE THESE AQUIFERS FOUND GLOBALLY?

The total surface area of Earth is about 510,082,000 km^2, give-or-take a few square centimetres. Oceans and seas cover about 71 % of the surface area. Indeed, the world's oceans/seas contain about 95 % of Earth's water (Table 1), excluding water in Earth's mantle. The total volume of water in these oceans and seas is approximately 1.35 billion cubic kilometres, or 1.35×10^{21} litres (that is, 1,350,000,000,000,000,000,000 litres), with an average depth of almost 3,700 metres. Unfortunately, seawater is highly saline (about 34 g of salt dissolved in each litre of water) and cannot be used for drinking, nor irrigation, without desalination.

Table 1: The global distribution of water, as a percentage of the *total* volume of water (saline plus freshwater).

Oceans	97.2 %
Ice Caps/Glaciers	2.0 %
Groundwater	0.62 %
Freshwater Lakes	0.009 %
Inland seas/salt lakes	0.008 %
Atmosphere	0.001 %
Rivers	0.0001 %
TOTAL	99.8381 %

The total volume of freshwater on Earth is about 34,650,000 km^3. Sounds a lot, but unfortunately most of this *freshwater* is not liquid, it is frozen. About 2 % of *all* Earth's water is frozen (Table 1) in two massive ice sheets, one covering Greenland and one covering Antarctica. Ice sheets are layers of frozen water sitting on land and should not be confused with the floating ice at the north pole (the Arctic). While only about 2 % of *all* water is in these ice sheets, ice sheets and glaciers account for *almost 70 % of all Earth's freshwater*. Worryingly (because scarcity of water is a globally important issue, affecting 2 – 4 billion people each year), only a tiny fraction of *all freshwater* is liquid on the surface of Earth – about 0.3 %, or about 93,000 km^3. Figure 1 illustrates one interpretation of the distribution of water among the various major pools (pun intended).

Because about 70 % of all *freshwater* is frozen and to all intents and purposes, not readily available, the volume of water readily available for human use is very small. However, there is a very large store of water underground in aquifers. Globally, a total of about 23,400,000 km^3 of GW may exist; about 2 % of all water on Earth. Unfortunately, it is extremely difficult to accurately quantify the volume of water in aquifers and this number is hugely rubbery. It is also unfortunate that about 50 % of this GW is saline, too saline for human consumption or use in irrigation.

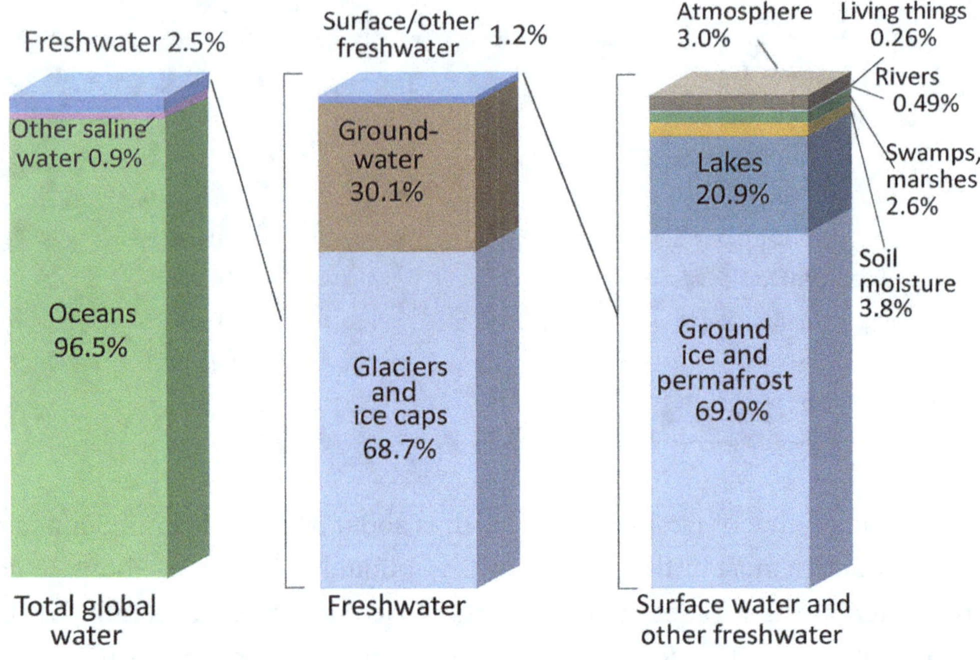

Figure 1: The distribution of Earth's water. (Image is public domain).

WHERE IN THE WORLD IS GROUNDWATER FOUND?

The location of the 37 largest aquifers is given in Figure 2. Every permanently-inhabited continent contains at least two large aquifers, and multiple smaller aquifers.

Many people consider the largest aquifer in the world to be the GAB in Australia, (it underlies about 1.7 million km^2, or about 25 % of Australia (Fig. 3), and is about 7 times larger than the area of the UK). While the GAB is the deepest aquifer in the world, the West Siberian Artesian Basin is larger, in terms of area of land sitting above, underlying about 2.703 million km^2 of EuroAsia. It certainly supports a far larger human population (about 13 million) than the GAB (fewer than 250,000 people). The Amazon Basin (I am, of course, referring to the aquifer called the Amazon Basin, not that part of South America that is drained by the

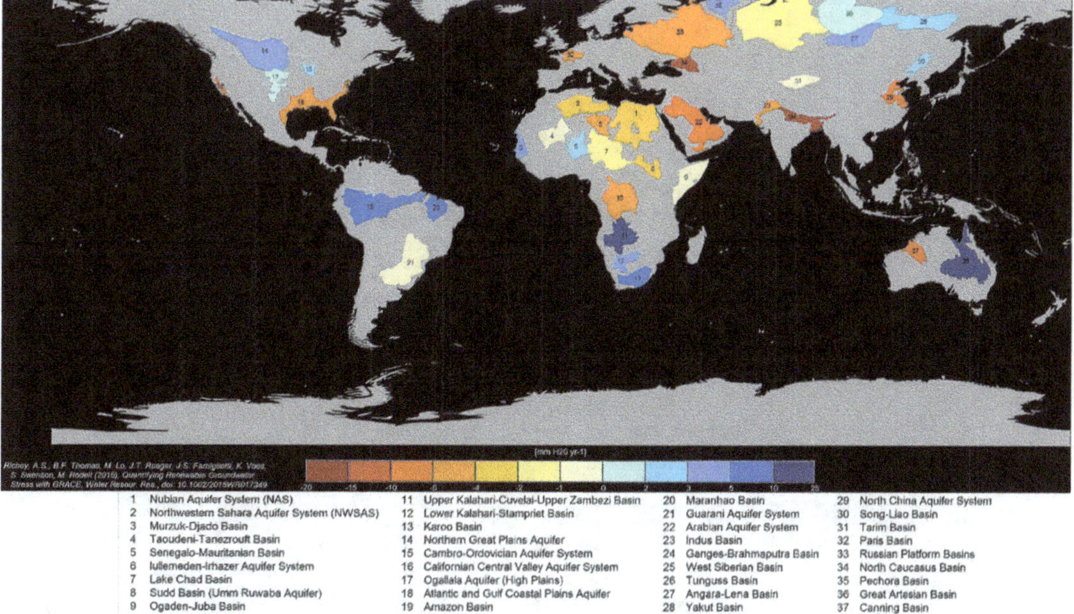

1 Nubian Aquifer System (NAS)	11 Upper Kalahari-Cuvelai-Upper Zambezi Basin
2 Northwestern Sahara Aquifer System (NWSAS)	12 Lower Kalahari-Stampriet Basin
3 Murzuk-Djado Basin	13 Karoo Basin
4 Taoudeni-Tanezrouft Basin	14 Northern Great Plains Aquifer
5 Senegalo-Mauritanian Basin	15 Cambro-Ordovician Aquifer System
6 Iullemeden-Irhazer Aquifer System	16 Californian Central Valley Aquifer System
7 Lake Chad Basin	17 Ogallala Aquifer (High Plains)
8 Sudd Basin (Umm Ruwaba Aquifer)	18 Atlantic and Gulf Coastal Plains Aquifer
9 Ogaden-Juba Basin	19 Amazon Basin
10 Congo Basin	

20 Maranhao Basin	29 North China Aquifer System
21 Guarani Aquifer System	30 Song-Liao Basin
22 Arabian Aquifer System	31 Tarim Basin
23 Indus Basin	32 Paris Basin
24 Ganges-Brahmaputra Basin	33 Russian Platform Basins
25 West Siberian Basin	34 North Caucasus Basin
26 Tunguss Basin	35 Pechora Basin
27 Angara-Lena Basin	36 Great Artesian Basin
28 Yakut Basin	37 Canning Basin

Figure 2: Map of GW Storage in the Earth's 37 largest aquifers. Aquifers coloured red/orange are experiencing significant declines because extraction exceeds recharge. Aquifers coloured blue have recharge exceeding extraction. From: https://gracefo.jpl.nasa. gov › resources › map-of-groundwater. Image credit: NASA/JPL-Caltech.

Amazon River and its tributaries) in South America is also larger in areal extent (2.28 million km^2) and the number of people accessing it (about 9 million people). The Ogallala Aquifer is the largest aquifer in the United States. It is part of the High Plains aquifer system, which underlies parts of eight states from Texas to South Dakota. It underlies an area of about 450,000 km^2, supports the drinking needs of more than 2 million people, and more than 25 % of all irrigated land in the US sits above this aquifer.

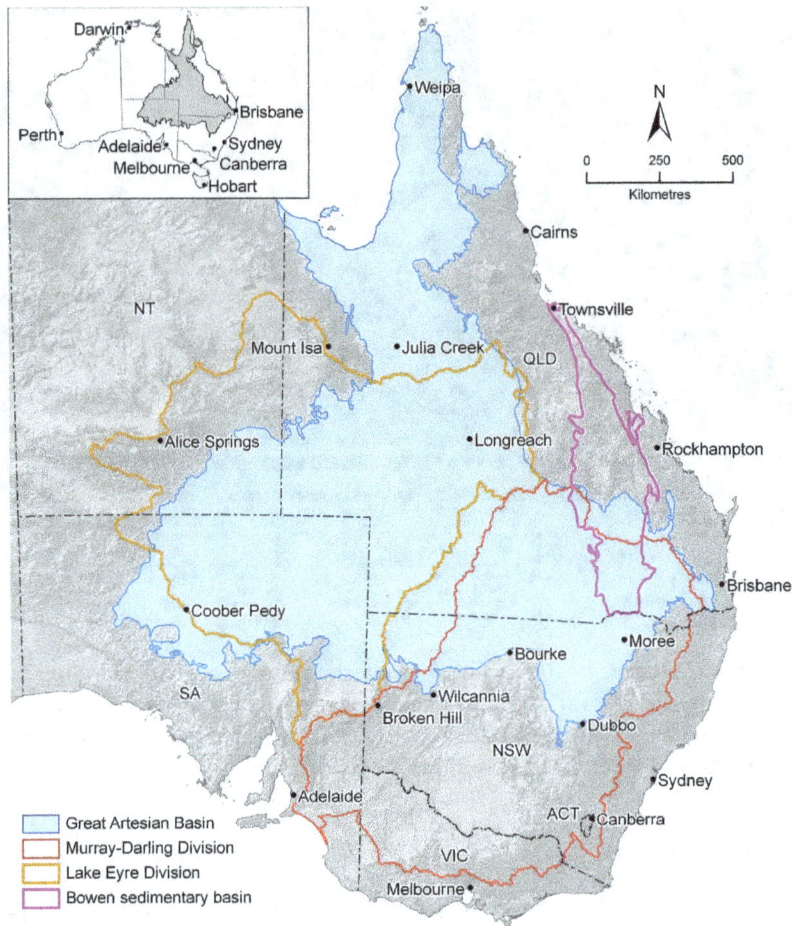

Figure 3: The distribution of the Great Artesian Basin in eastern Australia. From: Image from: https://www.environmentbuddy.com/news/facts-about-the-great-artesian-basin/ Image credit CSIRO 2011.

WHERE DOES GROUNDWATER COME FROM? THE GLOBAL WATER CYCLE

Groundwater is recharged (refilled) from multiple sources, including rainwater, meltwater, rivers/wetlands (losing streams and losing wetlands; see below), and artificial recharge. The best way to think about this is to consider the global water cycle (Fig. 4).

Rain and snow fall on the land surface (I have ignored rainfall to the oceans merely for convenience, obviously this is a major inflow of fresh water to oceans). Rain seeping into soil has three fates. It can be used by plants for transpiration,

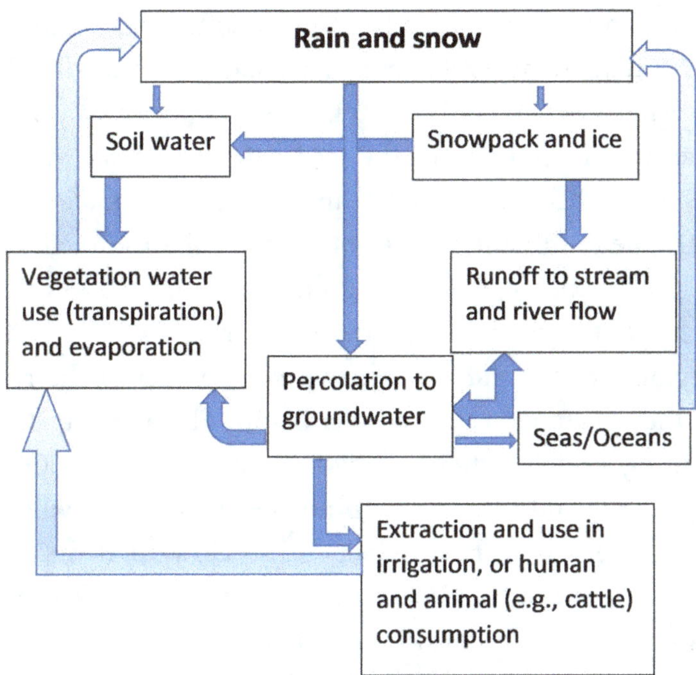

Figure 4: Some of the major flows of water in the water cycle. The solid dark blue arrows indicate liquid or solid (snow) fluxes of water, the light blue gradient arrows represent vapor flows as transpiration or evaporation.

thereby removing water from the soil and returning it to the atmosphere, where it can, at some point, form clouds, to later fall, usually at a different location from where it was transpired (although about 50 % of transpired water from the Amazon falls, again, somewhere within the Amazon basin). Alternatively, rainwater can move across a landscape, either as surface lateral flow or as lateral flow within the soil profile, into rivers and streams, which generally flow to the ocean (or a sea). The exception is in endorheic basins, where rivers and streams don't reach the ocean (see text box below) and instead water is retained in lakes and swamps within the basin. Rainwater can otherwise percolate deeper through the soil profile and enter an aquifer, that is, the store of GW that occurs beneath some, but not all, landscapes. Thus, *one source of recharge of GW is recent rainfall.*

In some catchments around the world, snow and ice accumulate in winter and melt during the following summer. Annual glacial meltwater, and spring meltwater from snow and ice on mountains, are very significant water resources globally, with more than 2 billion people reliant on such resources for drinking, irrigation, and stock watering. Ice and snow accumulated in mountainous regions in winter acts not only to store water, but also releases water relatively gradually. The Indus River, with an annual water flow exceeding 240 km^3, supports a vast population and has supported numerous civilisations in the past. It is the mainstay of agriculture in Pakistan. Meltwater from glaciers, snow, and ice in the Himalayas and the Hindu Kush of Tibet are the dominant sources of water in the Indus. It is this meltwater that dominates GW recharge, accounting for more than 80 % of recharge in the cold upper reaches of the river. Much of the recharge of GW in this region occurs from "losing regions" of the Indus river, that is, stretches of the river where river water is lost via the streambed to the aquifer below. The opposite of a losing stretch of river is a "gaining stretch", where a length of river intersects an aquifer and the aquifer discharges water into the river. In northern Australia, where a 5–6-month dry season occurs every year, rivers such as the Howard River continue to flow all year because of the inflow of GW during the dry season. Wetlands can also be gaining and losing wetlands according to whether they are receiving GW inflows or whether water from the wetland is recharging the aquifer below.

In Europe, the Alps are the largest suppliers of meltwater, while in North America, the Rocky Mountains of Colorado supply water to the Colorado River and the Rio Grande (among many others) and contribute significantly to GW recharge.

Thus, meltwater is a significant source of groundwater recharge on every permanently inhabited continent.

Finally, artificial recharge occurs in many regions globally, but especially in India, Pakistan, and China, where excessive extraction has caused the water table to fall hundreds of metres. Artificial recharge uses man-made structures, including infiltration basins and injection wells, to increase the rate of water inflow to an aquifer, thereby increasing recharge.

Endorheic basins and lakes

An endorheic (also spelled endoreic or endorreic) basin is a drainage basin that normally does not support outflow to other external bodies of water, such as rivers or oceans. Any flow of water that does occur within the basin converges into lakes or swamps, which can be permanent or seasonal. Some rivers are endorheic and these do not flow to an ocean (or sea). Examples of endorheic rivers that flow into (endorheic) lakes include the Jordan River, which flows into the Dead Sea (which is not a sea but a hypersaline lake), the Kabera river flowing into Lake Victoria (in sub-Saharan east Africa). Lake Eyre Basin in Australia is the largest endorheic basin in Australia and one of the largest in the world, covering about 1.2 million km^2, including much of inland Queensland, a large fraction of South Australia and the Northern Territory, and a small part of western New South Wales. Several rivers flow, occasionally, into Lake Eyre (itself an endorheic lake), but often the rivers dry out before reaching the lake. The lake itself is generally dry, filling up only 2 or 3 times each century.

WHAT IS GROUNDWATER USED FOR?

Groundwater has been used (unknowingly) by humans for as long as humans have existed, simply because many marshes, streams/rivers, and lakes are fed by GW. Oases and springs are also dependent on GW. The discharge of GW to the surface of Earth allows ready access to GW, although it hasn't always been apparent to humans that GW was being used when the source of the GW was a gaining stream/ river or lake.

As previously mentioned, hominins and early humans used springs and spas to support them, and access to such water resources (and the flora and fauna associated with them) was central to survival and migration out of Africa. Humans have since become increasingly sophisticated in accessing GW. Indeed, there is a 3000-year-old technology that is remarkable for its engineering prowess, sophistication, and understanding of hydrology, and reflects the ability to mobilise society to achieve a collective end. A qanat, also called a kariz or foggaras, is an *underground* system of canals that moves water from an aquifer to the land surface (Fig. 5). This technology was developed in Iran and Iraq and may then have spread west and east to Europe and China, among other countries, although some have suggested it was developed independently in South America and China. Some qanats remain in use to-day.

By moving water underground, at the depth of the water table, to lower elevations sites where GW intersects the ground surface (or is very close to the ground surface), large amounts of human effort is not required to move water up to the land surface. Additionally, by keeping water below-ground, rates of evaporation are minimal because the water surface is not exposed to hot dry air and solar radiation. Water supplies are also almost independent of variation in local rainfall. Although mostly used for irrigation and stock and human consumption, the cooling effect of evaporation of water was also used to cool air passing through a qanat. This cold air was then fed into a large, thick-walled building containing a tall ventilation chimney. Ice (taken from nearby mountains) was stored throughout the summer months within these buildings, in Persia (see Chapter 4).

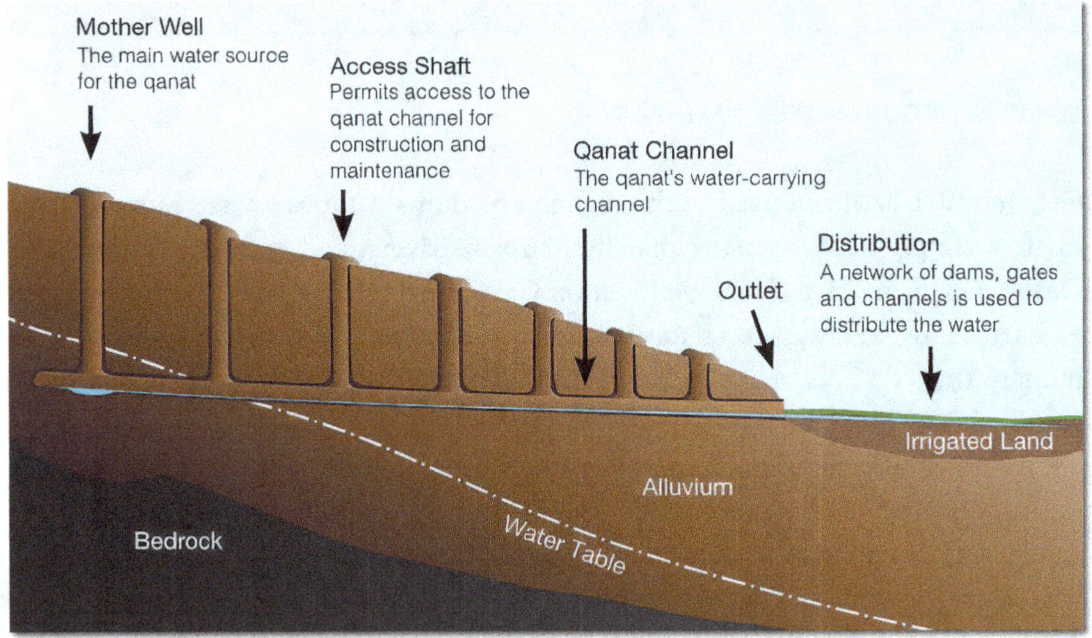

Figure 5: A schematic illustrating the principal features of a qanat. From Wikipedia (https://en.wikipedia.org/wiki/Qanat; original artwork by Samuel Bailey, 2009). Image created under Creative Commons Attribution 3.0 Unported license.

In summary, the importance of GW can be summarised by the following statistics: it is estimated that, globally, about 36 % of drinking water is derived from GW, about 42 % of irrigation water is derived from GW, and 24 % of industrial water use is derived from GW.

The analysis above is anthropocentric, focusing on the uses of GW that are *directly* linked to humans. However, such a focus is flawed because it ignores the central importance of GW to the ecology of numerous landscapes/ecosystems, which I now discuss.

GROUNDWATER DEPENDENT ECOSYSTEMS

It is not generally realised that a very large number of ecosystems, globally, are reliant on a supply of GW. Three classes of GW dependent ecosystems (GDEs) can be readily identified:

I. Ecosystems relying on the surface expression of GW, for example wetlands, springs, streams/rivers, and lakes that receive GW directly. The flora, fauna, and microbial populations of these surface water expressions of GW are dependent on the input of GW. The Great Artesian Basin of Australia produces a very large number of springs, called Mound Springs. These surface expressions of GW were vital, first to Aboriginal life and culture, and then to the European settlers of the 18th century, facilitating migration and economic development of vast areas of arid Australia. Mound Springs in Australia are associated with many rare endemic species, including invertebrates, flowering plants, and fish.

II. Ecosystems that access GW via the roots of vegetation, including riparian forests (forest growing along the sides of streams and rivers) and some woodlands and forests (Fig. 6). The iconic River Red Gum (*Eucalyptus camaldulensis*), is found extensively across Australia and frequently accesses GW, allowing it to survive in, for example, the arid interior of the continent. Similarly, in California, numerous phreatophytic (accessing GW via their roots) riparian tree species have been identified, including *Salix gooddingii* (willow), *Populus fremontii* (cottonwood), and *Quercus lobata* (oak). Across every continent, woodland and forest tree species reliant on GW have been identified.

III. Aquifer and cave ecosystems where stygofauna (animals living within GW, including crustacea, worms, bacteria, and fish) are present. Such stygofauna are generally colourless, blind, have very low metabolic rates, are long-lived, and locally endemic with a very restricted distribution. Protecting such species is fraught with difficulties because of the importance of GW extraction to local and regional economies and human life.

Figure 6: A tall dense eucalypt woodland growing above a 2 m deep aquifer in south-eastern Australia. The diameter and height of the trees and their growth rates are significantly enhanced during dry periods compared to nearby woodlands where GW depth exceeds 30 m.

Given the global distribution of GDEs it is surprising to realise that they have only received significant scientific and policy interest over the past 60 years, or so.

ARE WE EXTRACTING TOO MUCH GROUNDWATER, AND WHAT HAPPENS WHEN EXTRACTION EXCEEDS RECHARGE?

If we go back 10,000 years, it is reasonable to assume that GW stores were in equilibrium, such that the rate of recharge to an aquifer was the same as the rate of discharge from that aquifer. Discharge was occurring through transpiration of vegetation and discharge to oceans, rivers, wetlands, springs, and lakes. Even 3000 – 6000 years ago, when wells had been dug, qanats (and their regional equivalents) constructed, canals constructed, and irrigation as a technique was spreading, the human impact of GW extraction on GW stores was undoubtedly minimal. But the advent of industrialization allowed far larger volumes of GW to be extracted and be extracted from larger depths. Extraction gradually became unsustainable, partly because it was assumed for many years that the sustainable rate of extraction was equal to the rate of recharge. However, this is clearly not true because as extraction increased, discharge to wetlands, streams/rivers and lakes declined and the depth of the water table increased to depths that made it too deep for access by roots of GW dependent vegetation.

Over-extraction (unsustainable rates of extraction) of GW has been documented extensively, including from the Great Artesian Basin, the High Plains aquifer in the USA, north-western India (India is the largest user of GW, globally; 75 % of aquifers in Punjab, for example, are over-exploited), northern China, the Middle East, and north Africa. The ratio of annual GW extraction to estimated rates of recharge exceeds 3.5 in Egypt, is about 8 in Libya, and 9.5 in Saudi Arabia, indicating a very large rate of unsustainable extraction.

There are three clear impacts of over-extraction of GW. These are: (i) negative impacts on the health and ecology of GDEs. For example, the rate of growth of trees that are accessing GW declines rapidly when access to GW is reduced or removed, GW-fed wetlands rapidly dry out and disappear when the water table sinks a couple of metres, river flow in the dry season may stop and rivers dry out, when GW influx declines to zero; (ii) social and economic impacts as GW for human consumptive use dries up, GW for stock and crop irrigation becomes insufficiently available or entirely unavailable, and the viability of major industries, such as the Olympic Dam mine, are threatened if access to GW becomes limited. The Olympic Dam mine is a large underground mine in South Australia, and globally, is the fourth largest copper deposit and the largest known deposit of uranium. It currently uses about 35 ML of GW per day (almost 13,000 ML annually); (iii) structural impacts on infrastructure. Examples of structural impacts are given in the following two sub-sections. The first discusses the problems being experienced by Jakarta, in Indonesia; the second discusses subsidence of the North China Plain.

JAKARTA IS SINKING DUE TO EXCESSIVE GW EXTRACTION

Jakarta, the capital of Indonesia, is a very large city, with a metropolitan population of about 35 million when its satellite cities (for example, Depok and Bogor) are included, making it the second most populous city in the world (after Tokyo). It is a low-lying coastal city sitting above an upper unconfined aquifer at a depth of less than 40 m, a middle aquifer at depths of 40 and 140 m, and a lower aquifer at depths of 140 and 250 m. Extraction of GW from these aquifers resulted in subsidence ranging from 20 – 200 cm across Jakarta between 1982 and 1997. Continuing subsidence is largest in the northern suburbs, close to the sea. The depth of the piezometric head (the height that water pressure within an aquifer can push water up from an aquifer) of the middle aquifer increased from about 35 m below sea-level to about 54 m below sea level between 1989 and 1995, associated with subsidence across the city. Increased depths of piezometric depth means that the water table is moving away from the ground surface. During particularly high tides, flooding of coastal suburbs is increasingly common, and the depth of flooding is increasing, as the land surface subsides. So – what is being done in response to this?

In 2021 the Parliament of Indonesia passed a Bill authorising the construction of a new capital, to be called Nusantara, at a site in the jungle of Kalimantan on Borneo Island. Moving the capital has been proposed many times in the past 50 years because of pollution and flooding problems. Moving the capital is going to be hugely expensive, and is mostly very unpopular.

CHINA IS EXPERIENCING EXTENSIVE SUBSIDENCE ACROSS BEIJING AND THE NORTH CHINA PLAIN

More than 50 major cities in China, including Beijing and Shanghai, suffer from land subsidence. This is because the water table has fallen by up to 90 m at some wells across the North China Plain (NCP). A 2009 survey showed that 79,000 km^2 of China's land surface has already sunk by more than 20 cm. The Beijing Plain (located NW of the NCP, see below) is made up of multiple alluvial fans (a triangular shaped accumulation of sediments (e.g., silt, sand) emerging from a concentrated source of sediments (e.g., a narrow canyon through which water flows periodically in a wet season)). In the past 50 years, subsidence of up to 140 mm y^{-1}

has been recorded across the Beijing Plain. A total subsidence of 600 mm has resulted in major problems for the operation of the high-speed (350 km h^{-1}) railway between Beijing and Tianjin, while extensive cracking of buildings, roads, underground pipes, and airports has been recorded. Pumping of 44.8 km^3 of water, or almost 0.05 % of all liquid freshwater on Earth, annually from the Yangtze River to northern China is being undertaken as a means of reducing GW extraction while simultaneously increasing the areal extent of irrigated crops in northern China. This is further discussed several pages below. About 120,000 km^2 of the NCP (almost 90 % of the total area of the plain) has experienced significant subsidence, with falls of almost 3 m recorded in the worst affected regions. Rates of subsidence across the NCP range from about 5 mm y^{-1} to about 35 mm y^{-1}, between 1970 and 2015, although droughts can increase these rates by a factor of two because of reduced recharge by rainwater and increased extraction to compensate for lack of rainfall. Significant negative impacts on infrastructure, the economy, and society are being experienced across the entire NCP. Let us take a look at the Chinese government's response to overextraction in the north.

CHINA'S PAN-CONTINENTAL RESPONSE TO REGIONAL WATER SHORTAGE: THE GREAT CHINESE S-N WATER TRANSFER

China has a population of about 1.4 billion people, scattered across about 9.6 million km^2. Of this total land area, about 3.7 million km^2, or 38 %, has an arid, semi-arid, and dry sub-humid climate, meaning that monthly and annual rainfall is much less than the potential rate of evaporation. The potential rate of evaporation is the depth of water that could be evaporated if there was this much surface water available. But there isn't, so actual evaporation is less than potential evaporation. Potential evaporation is determined by the amount of energy available to evaporate water at a site. When the regional potential rate of evaporation is much larger than the actual rate of evaporation in a region, water resources are scarce and reliance on GW is generally high.

Feeding 1.4 billion people is no simple task and the NCP is probably the most important grain-growing region in China. The plain covers an area of about 409,500 km^2 (Fig. 7). The yellow soil of the plain supports vast areas of sorghum, wheat, sesame, millet, maize, and cotton production. The plain is one of the most densely populated regions in the world (population of the NCP is about 220 million, or 500+ people per km², but rainfall is low, ranging from about 400 to about 1400 mm y⁻¹. Sustaining and growing the yield of crops of the NCP has required extensive use of water, including an ever-increasing use of GW.

The NCP overlays one shallow unconfined aquifer and three deep confined aquifers at depths of 40–60 m, 120–170 m, 250–350 m, and 400–600 m. The aquifers consist of sand, gravel, clay, and silt. Since the 1970s, rapid agricultural and industrial development has resulted in massively increased demand for water resources. Most of the major rivers in North China, including the Hai River, are dammed for hydroelectric power generation and urban water use. The Yellow River flows through the western and southern parts of the region but only a small percentage of surface water is used for irrigation, partly because the river is heavily polluted.

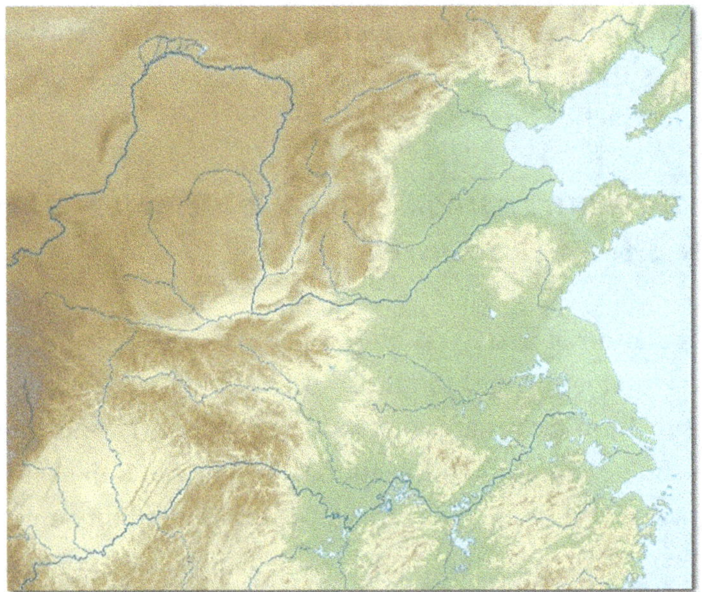

Figure 7:   The North China Plain (shaded green). This file is made available under the Creative Commons CC0 1.0 Universal Public Domain Dedication. Author: Kanguole.

Groundwater extraction has occurred in the NCP for a very long time, but the increase in pumping over the past 60 years has increased the depth to the water table by as much as 50 – 100 m, making it increasingly difficult and expensive to extract. Generally, GW quality declines with increasing depth, and increasing levels of salinity are being recorded in the aquifer. Salt is accumulating in the upper soil profile because salty GW it is being used to irrigate crops. Such accumulation of salt reduces soil quality and crop yields. By 2004, there were more than 7.6 million bores (or wells) in North China, and 68 % of irrigation was from GW. In 2009, GW accounted for 61 % of the total water supply to Beijing, 26 % in Tianjin, 80 % in Hebei province, and 58 % in Shanxi province. Large cones of depression occur with continued GW extraction (a serious problem for pumped wells because the water table height is increasingly depressed around the pumping bore). Over extraction of GW has resulted in severe GW depletion across the NCP, as evidenced by both bore data and two GRACE studies (see below for description of a GRACE study: it uses a pair of satellites orbiting Earth to measure small changes in the gravity field around Earth). As discussed earlier, severe GW extraction is also associated with land subsidence, an increasingly common problem in northern China. Increased desertification of northern China is also a major problem associated with the loss of ground cover by native vegetation due to over-extraction of GW. Desertification has been estimated to affect more than 30 percent of the total area (approximately 3.327 million km^2) of China, adversely affecting 400 million people.

So what is China doing in response to the ever-increasing problem of the depletion of GW in a region that has been responsible, to a large part, in feeding China?

Southern China gets quite a lot of rain each year, in the range 1000 – 1800 mm y^{-1}. Unfortunately, it is about 5000 km from southern China to northern China. However, 50 years after the idea was first suggested, by Chairman Mao Zedong, a massive project, the 'South-to-North Water Diversion Project', was initiated two decades ago. This project involves moving water from southern rivers to the dry north through a number of canals (Fig. 8).

This massive project began in 2002 and could be completed by 2050 at a cost of about $62 billion. It will eventually divert 44.8 x 10^{12} litres (or 44.8 km^3, or almost 0.05 % of all liquid freshwater on Earth) of water annually from the Yangtze River to northern China and will link China's four main rivers, the Yangtze, Yellow River, Huaihe and Haihe, and requires the construction of three diversion routes, stretching south-to-north across the eastern, central, and western regions of the country. Of course, it remains to be seen what the diversion of almost 45 km^3 of water from southern to northern China will do to the climate and ecology of southern and northern China. The evaporation of water caused by sunlight (called latent heat flux by micrometeorologists) is a major component of the energy balance of

Figure 8: The starting point of the Central Route of the South–North Water Transfer Project. Looking "upstream", toward the Danjiangkou Reservoir, from which the water is coming. Picture from Wikipages Creative Commons License. https://commons. wikimedia. org/ wiki/File:South_E2_80_93North_Water_Transfer_Project_Central_route_starting_ point_taocha.jpg

a region. One of the reasons that droughts are associated with higher land surface temperatures is because the absence of soil moisture and the absence of transpiration from leaves (because of the lack of soil moisture) converts an increasingly large proportion of incoming solar radiation into sensible heat flux, also known, and felt, as "heat".

SATELLITES, GROUNDWATER, AND GRAVITY

In 2002, two satellites were launched from a Russian site in a joint venture with NASA and the German Aerospace Centre. These two satellites, called GRACE-1 and GRACE-2 (GRACE stands for Gravity Recovery and Climate Experiment) accurately measure variation in the gravity field around Earth. The movement of tides and ocean currents, the seasonal and inter-annual changes in the volume (and hence mass) of water in very large aquifers, changes in the thickness of ice sheets, and even magma flows, can be detected as changes in the gravity field above the Earth's surface (Fig. 9). The majority (but not all) of the changes in gravity around the world are the result of changes in the total mass of water within a region. Annual patterns of water resources in the Amazon and changes in aquifer storage have been detected using these satellites. But how do they do this?

The two satellites are 220 kilometres apart, 500 kilometres above Earth. Microwaves emitted from each satellites to the other are used to measure the distance between the two satellites, to an accuracy of about 10 microns. When one satellite travels over a region of increased gravity, it accelerates a little and moves ahead of the other. This increase in separation is detected and recorded. As this satellite moves away from the region of increased gravity, it slows down and the second trailing satellite experiences the gravity anomaly and accelerates a little and catches up with the first satellite. Again, this change in separation is measured and recorded. Across a 30-day period the two satellites can map gravity anomalies across the entire globe.

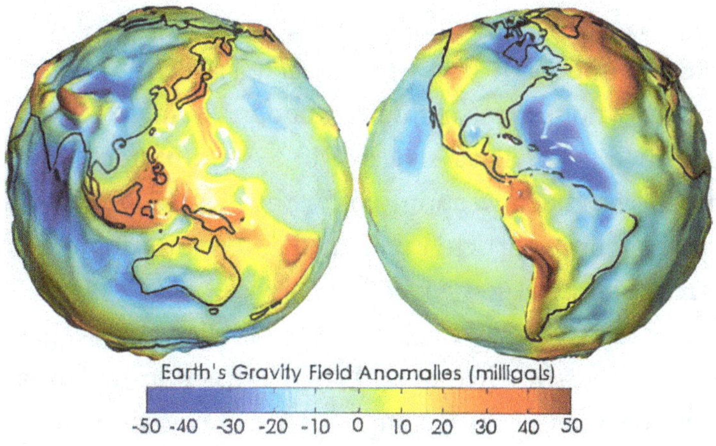

Figure 9:   Two "*gravity anomaly*" maps showing where models of the Earth's gravity field based on GRACE data differ from a simple mathematical model that assumes Earth is perfectly smooth and featureless. Areas colored yellow, orange, or red are areas where the gravity field is larger than predicted by the model (e.g., the Himalayan mountains in Central Asia). Progressively darker shades of blue indicate places where the gravity field is less than predicted by the Earth model. From: http://earthobservatory.nasa.gov/Features/ GRACE/page3.php

In 2010/2011 there was a massive rainfall anomaly in the southern hemisphere. For example, the long-term average rainfall at a research site of mine in Central Australia is 319 mm. In the 2010/2011 hydrological year (August 2010 to July 2011), more than 570 mm of rainfall was recorded. Indeed, Australia experienced one of the strongest La Niña episodes ever recorded. Consequently, rainfall across much of Australia was much larger than the long-term average. This had two important impacts on Australia. First, the volume of water held on the continent in soil, aquifers and lakes increased to such an extent that it was detected by the GRACE satellites. Second, much of this water was used by vegetation to support a very strong surge in growth and the continent "greened-up" which means the total amount of leaf area (and hence chlorophyll; hence the phrase "greening-up") sitting across the continent increased. This too was detected by other satellites that can measure the amount of chlorophyll and productivity of landscapes. My eddy covariance site (Chapter 10) detected the impact of this increase in rainfall in arid central Australia as a large increase in photosynthetic C uptake during 2010/2011.

In conclusion, I think we can summarise this chapter by saying the following:

Groundwater is a very large volume of water (mostly freshwater) residing underground in aquifers. It was a major factor affecting the evolution and migration of early hominins (including modern humans) when our ancestors were confined to the East African Rift System of north-eastern Africa.

Access to GW has been weaponised for the past 4500 years and multiple examples can be found, particularly in arid and semi-arid landscapes where surface water resources (lakes, rivers) are scarce.

Groundwater is replenished directly by rainfall, water seepage from rivers and streams, and snow/ice melt. When extraction by Humans exceeds recharge, GW resources are depleted. This has negative impacts on river flows, wetland ecology, availability for future consumptive use, and, perhaps most dramatically, land subsidence. Such subsidence is heavily impacting the city of Jakarta (the capital of Indonesia with a population of about 35 million) and subsidence of Beijing and agricultural land across the Beijing Plain and the North China Plain (population exceeding 340 million, including Beijing, which sits on the NCP). Not only is land subsiding across these plains, but GW is becoming more saline (making it unsuitable for agricultural use and human consumption) and it is becoming increasingly difficult to extract GW from ever-increasing depths.

One tool being used to quantify and track GW resources is a pair of satellites, called GRACE-1 and GRACE-2, which measure anomalies in the gravity field around Earth.

THE END

Chapter 6

What makes the Earth's magnetic field? Why do we care about our magnetic field, and does it ever flip North to South?

Sure, we all know Earth has a magnetic field, otherwise compasses wouldn't work, but how is the Earth's magnetic field generated? Is there a giant bar magnet within Earth? And does the strength of the magnetic field vary? Perhaps more importantly, has the Earth's magnetic field ever dropped to zero (or close to zero) and would it matter if it did? Are the magnetic north and south poles fixed in one location, or do they move? If they do move, could our N and S magnetic poles ever flip, so that our N magnetic pole becomes our S magnetic pole? Does life-on-Earth benefit from the magnetic field around Earth? Has the magnetic field affected the evolution of humans? These questions form the basis for this chapter. But first, I include a point of clarification about the number of N and S poles we do have……

There are three poles in each hemisphere of Earth

Yes, there are three poles in each hemisphere: the geographic north/south pole, the north/south magnetic pole, and the north/south geomagnetic pole. The geographic north pole is the point in the Northern Hemisphere where the Earth's axis of rotation meets the Earth's surface. The geographic North Pole is the northernmost point on Earth, lying diametrically opposite the South Pole. It defines geodetic latitude 90° North and at the North Pole, all directions point south. Conversely, the geographic south pole is the southernmost point on Earth, lying diametrically opposite the geographic Nouth Pole. It defines geodetic latitude 90° South and at this point, all directions point north.

The north magnetic pole is a point on the surface of Earth's northern hemisphere where Earth's magnetic field points vertically downward. At any one time, there is only one location where this occurs. However, the north magnetic pole moves slowly over the surface of Earth (see below).

The north geomagnetic pole is a theoretical construct. It represents one of the two antipodal (diametrically opposite sides on the circumference of a circle; the north and south geomagnetic poles are diametrically opposite each other) geomagnetic poles. If the Earth's magnetic field were a perfect dipole (like a bar magnet) then the axis of this theoretical bar magnet would intersect the Earth's surface at the geomagnetic poles. Because Earth's magnetic field isn't a perfect dipole, the place where the magnetic field is vertical occurs at a significant distance away from the geomagnetic poles (that is, at the magnetic poles).

In this chapter, we are only concerned about the magnetic north/south poles.

FLIPPING OF N-S MAGNETIC POLES

Earth's magnetic poles have repeatedly and *completely* reversed their orientation (N became S and *vice versa*) during the geologic past; this process is called *geomagnetic reversal.* The most recent reversal occurred 780,000 years ago (the Brunhes-Matuyama reversal). The timing of such reversals appears to be random, with 183 reversals recorded over the past 83 million years (an average of about once every 500,000 years). It may take about 7000 y for the reversal to be complete, but this differs across the 183 reversals and estimates vary, with one recent estimate of 22,000 years for the B-M reversal. A reversal may last for 2000 – 12,000 years. The periods of constant polarity (either "normal" or reversed), are known as a *chron.* This is in contrast to short-term partial, or incomplete, reversals that last only a few-to-several hundreds of years, called excursions, with up to 45° of change in the location of the geomagnetic pole. During an excursion, the magnetic field of the liquid outer core reverses, but that of the inner solid core does not. The most recent excursion is known as the Laschamps Excursion and occurred 41,000 – 42,000 years ago (see p113).

How do we know that the N and S magnetic poles have flipped so many times? There are three sources of data.

The first source of data is contained in the lava discharged onto the surface of Earth. Lava contains iron. As the lava cools and solidifies, the orientation of the magnetic field at the site of cooling is "locked-in" and recorded by the ferrimagnetic minerals in the solid lava. By measuring the orientation of the magnetic field produced by the lava it is possible to demonstrate polarity reversals over the geologic past. A second source of data can be found in the world's oceans, where ridges (for example, the Pacific-Antarctic ridge in the southern Pacific Ocean and the Reykjanes ridge in the northern Atlantic Ocean) occur where two tectonic plates meet. Here, new ocean floor is being produced and as it is produced, previously-produced sea floor moves away from the ridge. The direction of the magnetic field is recorded in the newly formed ocean floor and if the magnetic field is reversed, this reversal will be recorded relative to the field recorded in older and younger parts of the ocean floor. This information is recorded symmetrically on both sides of the ridge and the pattern observed around the Pacific-Antarctic ridge matches that around the Reykjanes ridge. When the pattern of direction of the magnetic field is plotted as a function of time into the past, a "bar code" becomes apparent (Fig. 1) and these bar codes demonstrate that the duration and frequency of reversals have been extremely variable. The Cretaceous Superchron lasted more than 40 million years (that is, there was not a reversal of the magnetic field for a period of 40 million years). Such superchrons are very rare, with only two events identified beyond doubt. A superchron is defined as a period of time exceeding 10 million years when polarity ("normal" or reversed) is constant.

A third source of data is ferrimagnetic minerals in consolidated sedimentary deposits. The act of compaction (consolidation) of iron-containing sediments locks-in the local magnetic field, which can then be measured in the laboratory.

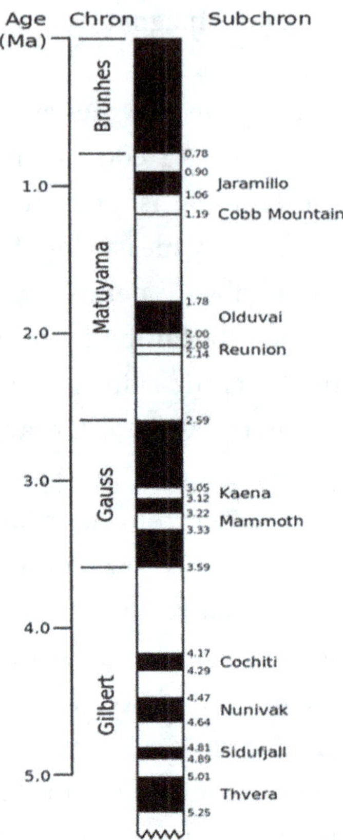

Figure 1: The "bar code" of the timing and duration of geomagnetic polarity over the past 5 million years. By United States Geological Survey, hand-traced to vector by a USGS staff member (User:Intgr) - U.S. Geological Survey Open-File Report 03-187, Public Domain, https://commons.wikimedia.org/w/index.php?curid=2878948

WHAT CAUSES THE POLES TO FLIP?

In simulations of the behaviour of the dynamo of the Earth's core, random and hence unpredictable reversals of the magnetic field are produced (Fig. 2). In laboratory experiments with molten metal, similar reversals have been observed. The Sun's magnetic field also flips, and flips extremely frequently – about once every 9 – 12 years. In contrast to the hypothesis that "flips are random and a trigger is not required to initiate a reversal" is the hypothesis that external triggers, such as impacts on the surface of Earth, or internal triggers, such as the movement of large slabs of mantle at subduction zones, may initiate reversals. Which is correct remains debated.

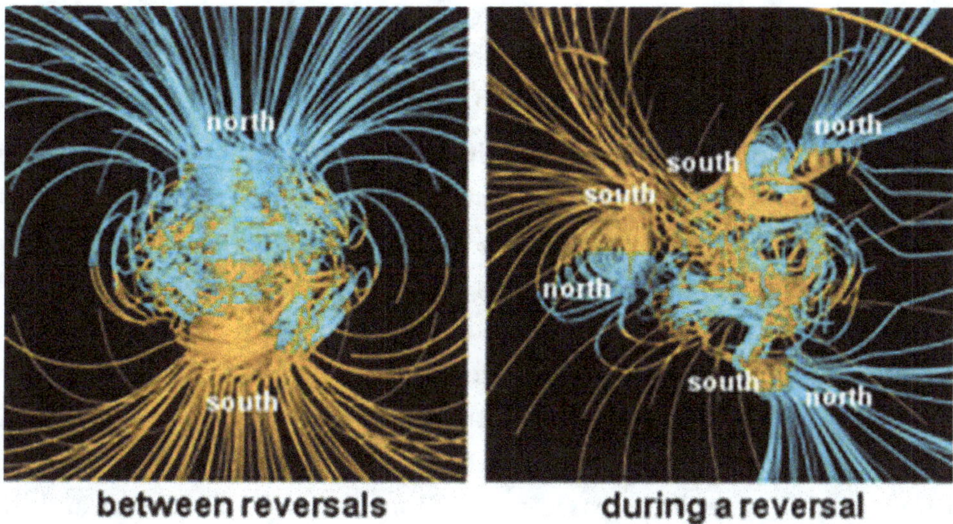

Figure 2: NASA computer simulation using the model of Glatzmaier and Roberts. The tubes represent magnetic field lines, blue when the field points towards the centre and yellow when away. The rotation of Earth is centred and vertical. The dense clusters of lines are within the Earth's core. Image from: https://en.wikipedia.org/wiki/Earth_27s_magnetic_field and is in the public domain, and was produced by NASA.

BIOLOGICAL IMPACTS OF POLE REVERSALS

In the past 170 years Earth's magnetic field strength has decreased by about 9 % and the speed at which the N geomagnetic pole is moving has increased significantly in the past 30 years. The last full pole flip occurred almost 1 Mya, but the most recent partial flip (where the poles change location by much less than 180°) occurred during the Laschamps Excursion (see below; the poles moved by about 45°), about 41,000 years ago. Are we due a (full or partial) flip soon? What could happen if the poles do flip?

Immediately preceding the Laschamps Excursion (41,000 to 42,000 years ago) the strength of Earth's magnetic field declined to less than 10 % of its usual intensity. During the short-lived period (about 500 years) of the partial reversal, the magnetic field strength was less than 20 % of its usual value. It is undoubtedly true that this large reduction in magnetic field strength will allow much larger levels of solar and cosmic radiation to reach the surface of Earth. Increased ionization and decreased stratospheric ozone concentrations have negative impacts on climate and biology – living cells do not cope well with receiving increased

UV, ionizing radiation and solar winds (see below). But is there any evidence for such impacts?

Climate proxy records of the Pacific Basin record local glacial maxima in Australasia and the Andes, long-term shifts in atmospheric circulation and continental-scale aridification and megafaunal extinction in Australia for the period 40,000 – 42,000 years ago. A recent study (Cooper and colleagues) used ^{14}C radiocarbon dating of ancient kauri (*Agathis australis*) trees preserved in wetlands of northern New Zealand to provide high-resolution dating during the Laschamps Excursion.

In their study, Cooper et al. (2021, Science) demonstrated that the combination of reduced magnetic flux strength and the grand solar minimum that occurred during the Laschamps Excursion resulted in large increases in atmospheric ionization from cosmic rays and increased production of oxides of nitrogen and hydrogen. These increases reduced atmospheric ozone levels, thereby increasing the level of UV radiation reaching the ground. Associated with these changes were changes in the temperature gradient between the equator and poles, a weakened Arctic polar vortex, and a net warming in the lower polar stratosphere. The authors speculate that the sudden change in behaviour of humans, with a large surge in cave dwelling, coincided with this period of increased UV radiation at ground level. Could the extinction of the Neanderthals that occurred at this time be linked to the Laschamps Excursion? Neanderthals and *Homo sapiens* coexisted in Europe for 5,000 – 10,000 years and interbred (we still have Neanderthal DNA in our DNA). Why did modern humans survive and Neanderthals die-out? While competition and aggression are possible reasons, the scale of inter-breeding was presumably large (because the DNA remains in our modern DNA), suggesting that perhaps competition and aggression wasn't the cause of the extinction of Neanderthals. A recent discovery of differences in the molecular structure of an intracellular chemosensor, the Aryl Hydrocarbon Receptor (AhR), between Neanderthals and modern humans, may be important. This chemosensor regulates gene expression, cell physiology, immunity, and cell differentiation. It is important in the response to skin cells to UV radiation. UV (especially UV-B, in the 290-320 nm waveband) is especially damaging to nuclear DNA. AhR contributes to DNA repair, tumour suppression, and skin tanning. The AhR of modern humans differs slightly from that of Neanderthals. It has been suggested that this difference gave modern humans an advantage over Neanderthals during the Laschamps Excursion, when the Earth's magnetic field was especially low and UV levels especially high.

A LONGER-TERM PERSPECTIVE OF DECLINES IN EARTH'S MAGNETIC FIELD AND EVOLUTION

The Earth's magnetic field helps maintain stratospheric ozone levels (by deflecting the solar wind before it can interact with the ozone layer). As such, it helps prevent harmful levels of UV radiation from reaching the surface of Earth. During periods of low magnetic field strength, rates of mutations in cellular DNA increase because of the increase in UV radiation at ground level (see the Laschamps Excursion discussed above).

Large declines in Earth's magnetic field strength have been recorded at 190,000 years ago (190 kya), 100-120 kya, 64 kya, 41 kya, and 13 kya. Interestingly, peaks in the rate of extinction of large animals occurred in the Americas at 13 kya, while Australia experienced large rates of extinction at 40 kya, 84 kya, 107 kya, and 122 kya. Eurasia and Africa experienced mass extinctions around 13 and 40 kya. Clearly, correlation is not proof of causation (Chapter 12), but this does seem a significant number of correlations between low magnetic field and mass extinctions (see text box) of large animals. But why are mass extinctions frequently confined to large animals (which are also long-lived, because it takes time to grow a large body mass)? Mutations to DNA are rarely beneficial. If you are large and long-lived, the probability of a deleterious mutation occurring somewhere in your body is larger than if you are small and short-lived. Consequently, large long-lived animals are at an evolutionary disadvantage when UV levels are high (i.e., when Earth's magnetic field is low). In addition, if you are small, burrowing, and curtail your activities to periods with the lowest UV (around sunrise and sunset), you can use these adaptations to survive. However, there are very few burrowing elephants in the world!

> **Mass extinctions**
> Mass extinctions are defined as the widespread and rapid (geologically speaking; about 3 – 5 million years) decline in the diversity of multicellular-species diversity. Five "big" mass extinctions can be identified, although arguments

exist about defining "big" and so there could have been more than 20 mass extinctions. Most extinctions occurred in the sea because this was where most species lived at the time. The earliest mass extinction – the Ordovician extinction – occurred about 445 Mya and the most recent mass extinction event occurred about 66 Mya, the K–Pg extinction event. However, in Australia mass extinctions started about 40,000 years ago, in Americas ~ 12 000 years ago, coinciding with early human migrations into these regions.

Multiple causes of mass extinctions are likely, including giant volcanic eruptions across thousands of years, large falls in sea-levels, asteroid impacts (the probable cause of the extinction of dinosaurs; Chapter 2), global cooling and geomagnetic reversals, among other possibilities. Many ecologists now refer to a 6th mass extinction event, the Holocene (or Anthropocene) extinction, which started somewhere between the mid-19th and early 20th centuries and continues to the present-day and is being caused by Human activities.

An alternative mechanism to explain mass extinctions and reversals in the magnetic field was proposed by Courtillot and Olson in 2007. They link the timing of mass extinctions to the end of superchron events, with a 10 – 20 million year delay between the end of the superchron and mass extinctions. They suggest that thermal instabilities within the mantle can increase rates of heat flow upwards towards the surface of Earth and this induces polarity reversals in the magnetic field, thereby ending the superchron (because the poles start to flip). But deep mantle plumes can move upwards through the mantle, taking about 20 million years to ascend. These plumes reach the Earth's surface and produce **massive** continental flood basalts. Flood basalts are the result of one or more gigantic volcanic eruptions that cover large stretches of land or the ocean floor with basalt lava. This **massive** volcanic activity can last thousands of years. The Deccan Traps (Chapter 2), for example, is a 2000 m thick layer of flood basalt that covers 500,000 km^2 in west-central India. It is the largest volcanic feature on Earth. It is this volcanic activity and the ash and SO_2 produced that cause mass extinctions by (a) cooling Earth (ash and SO_2 in the atmosphere stop solar radiation from warming Earth; Chapter 2); (b) acidifying oceans; and (c) spewing out molten lava that covers huge areas of land.

The timing of the next (full or partial) flip is impossible to predict. But the impacts on our technology are going to be severe, including a breakdown in satellite communications, navigation apps, and massive disruptions to power distribution systems, resulting in massive impacts on health, transport, communications, heating, cooling…, well, anything connected to a central electricity supply, really. Let's hope this doesn't happen tomorrow…..

EARTH'S MAGNETIC FIELD: HOW IS IT FORMED, AND IS IT IMPORTANT TO LIFE ON EARTH?

The magnetic field at the surface of Earth can be approximately represented as being produced by a giant bar magnet at the centre of Earth and tilted at an angle of about 11° off Earth's rotational axis. Earth's magnetic field originates within Earth itself and extends beyond the ionosphere to the edge of the magnetosphere (see below). But, the question remains: how is the magnetic field produced?

To answer the question posed above, we need to know just a little about Earth's internal physical structure.

There are essentially four layers to Earth: the crust, the mantle, and the outer and inner cores. The crust is the thin outermost layer upon which we reside. It is about 5 – 50 km thick (average about 15 – 20 km) but is thinner under oceans and thicker under the continents. Although continental crust is thicker than oceanic crust, it is less dense than oceanic crust, so it floats higher up on the mantle lying beneath the crust. This is fortunate as it allows the continents to remain above sea-level (mostly), supporting the development of forests, savannas, humans, and tigers (to pick a random selection of life-forms). The four most common elements in the Earth's crust are oxygen (nearly 47 %), silicon (nearly 28 %), aluminium (about 8 %), and iron (5 %). Oceanic crust is about 100 - 200 million years old, which is much younger than continental crust, which ranges in age from 1 – 4.3 billion years. Oceanic crust is much younger because of the occurrence of 'mid-ocean spreading centers' where new ocean crust is being formed, and this new crust moves out from the centre, pushing continental plates with them.

Earth's mantle is a layer immediately beneath the crust, sitting on the outer core. It is almost 3000 km thick (the radius of Earth ranges from about 6378 km at the equator and about 6357 km at the poles) and accounts for about 67 % of the Earth's mass, and about 84 % of the Earth's volume. Partial melting of the mantle at ocean ridges produces new oceanic crust. SiO_2 and MgO account for almost 84 % of the upper mantle, although these numbers are mere estimates, and the composition of the lower mantle remains speculative. The upper surface of the mantle (closer to the Earth's crust) is warm (about 200 °C) but the temperature of the lower surface (closer to the outer core) is hot (about 4000 °C). It is this temperature gradient that drives the circulation of mantle, with localised hotter regions of the lower mantle becoming more "molten" and less dense and so they upwell (ascend), while colder, more dense regions move down at convergent plate boundaries, called subduction zones.

The outer core lies at the centre of Earth and has an outer liquid layer, almost 2300 km thick, and an inner solid core with a radius of about 1200 km. The outer core has a temperature in the range of about 4000 – 6000 °C (the surface temperature of the sun is about 6000 °C) and is composed of iron (Fe) and nickel (Ni). The temperature of the outer core declines significantly from the innermost part to the outermost part. The inner core is also composed of Fe and Ni, and has a temperature of about 6000 °C.

But this doesn't tell us how the magnetic field is produced. It might be useful to remember that when a copper wire (doesn't have to be copper, any metal will do) carries an electrical current through it, the flow of electrons through the wire generates a magnetic field around the wire. The magnetic field can be thought of as being arranged as an infinite set of concentric circles that are perpendicular to the wire and the direction of the magnetic field lines (that is, their north-south polarities) can be detected with a compass. If wire is held in vertical orientation (straight up and down) and the **conventional** direction of the current is, let's say, from the lower to the upper end of the wire (because the wire is connected to a battery, with the upper end connected to the negative terminal and the lower end to the positive terminal) then the magnetic field lines can be viewed as being in an anti-clockwise direction, viewed from above the wire). This means that as a compass is moved in a circular path around the wire, the north pole of a compass will trace the magnetic field and rotate as the compass moves around the wire (Fig. 3). The strength of the magnetic field declines with increasing distance from the wire (just as gravity declines with increasing distance from the centre of a planet, although the former declines in a simple inverse relationship (1/D, where D is distance) the latter is an inverse square relationship ($1/D^2$)).

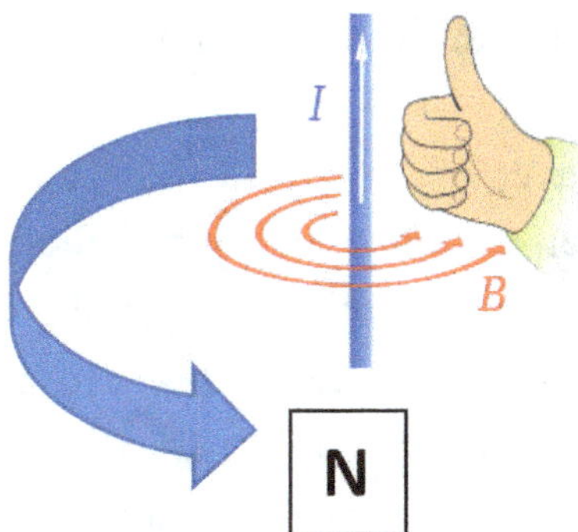

Figure 3. The image of the hand tells us that the thumb indicates the conventional direction of current flow along an electrical wire, and the orientation of the remaining fingers tells us the direction of the magnetic field around the wire – in this case, anticlockwise. The north pole of a compass would point to the right when placed in front of the wire but point to the left when placed behind the wire, as indicated by the wide blue arrow.

Image produced by Prof José Fernando, la Universidad de Granada. From https://en.wikipedia.org/wiki/Magnetic_field. This image is licensed under the Creative Commons Attribution-Share Alike Attribution-Share Alike 4.0 International, 3.0 Unported, 2.5 Generic, 2.0 Generic and 1.0 Generic license.

The key point here is that it is the *movement of charge* (electrons in the case of an electrical current in a wire) that results in the generation of the magnetic field. How this happens is not known. If you are wondering about a bar magnet, where electrical currents are not flowing because the bar magnet is sitting in the palm of your hand, and not attached to a battery, we must remember that all atoms, including the iron in a bar magnet, contain electrons. Electrons are negatively charged and also have angular momentum, referred to as spin. Generally, electrons like to be paired, with one electron of the pair having an "upward" spin and the other of the pair having a "downward" spin. The two spins cancel each other out and do not generate a magnetic field. Iron, however, contains four unpaired electrons and these can be aligned and their spins do not cancel each other out, thereby producing sufficient magnetic field that it can be experienced outside of the bar magnet. Nickel and cobalt are common metals added to iron to make permanent magnets because pure iron can't be used to make permanent magnets. Note that Earth's core is composed mostly of Fe and Ni. A permanent magnet is one that retains its magnetic properties in the absence of an electrical current flowing through the magnet. Electro-magnets are only magnetic when an electrical current is flowing through the wires that surround an iron core. The iron core is magnetic only when the electrical current is flowing.

Because the outermost layer of the liquid outer core is cooler than the innermost layer of the liquid outer core, the highly conducting Fe and Ni display convection currents, moving heat from the inner core towards the mantle, via the outer core. Because Earth is rotating, the Coriolis force (Chapter 2) makes these convection currents spiral in an anti-clockwise direction in the southern hemisphere and a clockwise direction in the northern hemisphere (in the same way that very large-scale ocean and wind currents rotate in opposite directions in the northern and southern hemisphere). We now have a *reverse* dynamo lying at the centre of Earth, in the sense that convection currents of fluid metal in the Earth's outer core, driven by heat flow from the inner core, are organised into rolls by the Coriolis force, which generate *circulating electric currents*, which supports the formation of a magnetic field. A dynamo has a coil of copper wire moving through a fixed magnetic field to produce electricity. The reverse can occur, whereby an electrical current flowing through a wire generates a magnetic field around it (Fig. 3). Within Earth, we have an electrical current flowing in the liquid outer core and this produces the magnetic field, which we can detect on the Earth's surface with a compass.

THE IMPORTANCE OF EARTH'S MAGNETIC FIELD TO LIFE ON EARTH

Sitting above the Earth's surface is, of course, the atmosphere, composed of several layers. The lowest layer is the troposphere, which contains O_2, N_2, CO_2, water vapor, and numerous other gases and dust. This layer extends about 12 km upwards. Almost all the weather we experience arises in the troposphere and almost all commercial flights occur within the troposphere. Above the troposphere lies the stratosphere (12 – 50 km above Earth's surface), where the ozone layer resides, which absorbs UV emitted from the Sun. Perhaps surprisingly, the temperature of the stratosphere increases as distance from Earth increases, because of this UV radiation. Above the stratosphere is the mesosphere (50 – 80 km above Earth's surface), which gets colder as distance from Earth increases. At the outer edge, temperature is about -85 °C. The thermosphere (80 – 700 km above Earth's surface) and the exosphere (700 – 10,000 km above Earth's surface) are the final two layers of the complete atmosphere. The lower part of the thermosphere contains the ionosphere and the *Aurora Borealis* and *Aurora Australis* can be seen in these layers when conditions are right. The International Space Station orbits in the thermosphere. The exosphere is the last layer of Earth's atmosphere and the solar wind (see below) starts to interact with the atmosphere here (hence the *Aurora Borealis* and *Aurora Australis*) (Fig. 4).

Figure 4: Northern lights (*Aurora borealis*) over the Víkurkirkja church at Vík í Mýrdal, Iceland. This picture was a finalist in Picture of the Year 2022 and is a 'Featured picture' on Wikimedia Commons and is considered one of the finest images in Wikimedia. Photo was taken by AstroAnthony. From: https://commons.wikimedia.org/wiki/File:Church_of_light.jpg

The ionosphere (50 km – 965 km up) is ionised by solar radiation (principally UV) and represents the inner edge of the magnetosphere. It consists of a shell of electrons and charged atoms and molecules and the amount of ionisation varies on daily, seasonal, and sunspot activity (approximately 11-year cycles) timescales.

It is the magnetosphere that is of interest to us in this chapter. A magnetosphere is the region of space around a planet (or a star) that is influenced by the planet's (or star's) magnetic field. It has a complex structure and the one around Earth plays a pivotal role in protecting life from the solar wind. The solar wind is a vast stream of charged particles released from the Sun's atmosphere (the Corona), known as a plasma. This plasma is mostly electrons, protons, and alpha particles (or alpha rays, namely, 2 protons plus 2 neutrons, identical to a Helium-4 nucleus), but atomic nuclei of Mg, Fe, Cr and Ni, among others, are also present. When the solar wind interacts with the magnetosphere, the *Aurora Borealis* and *Aurora Australis* can be formed.

> **The solar wind and comet tails**
> Comets are snowballs of frozen gases, rock, and dust orbiting the Sun. When frozen, they range in size from a few kilometres to tens of kilometres in diameter. However, when a comet's orbit brings it close to the Sun, solar energy warms the surface of the comet and dust and gases are boiled off into space, forming two tails – one composed of dust, the other of gases. The direction of the tail that we can sometimes see is not determined by the direction of travel of the comet. The tail is not the equivalent of the wake behind a ship sailing on a lake. The tail always points away from the sun because it is the solar wind and the pressure exerted by solar radiation that determines the direction of the tail.

In addition to the solar wind, galactic cosmic rays (GCRs) also bombard Earth. GCRs are high-energy protons (about 90 % of all GCRs are protons), plus the nuclei of all elements found on Earth (so they are composed of atoms that have had their electrons removed). GCRs originate outside our solar system (but our Sun emits cosmic rays, they just aren't called galactic cosmic rays) and may be produced when stars explode (supernova). Because GCRs are ions (being positively charged) they interact with and are influenced by Earth's magnetic field. When GCRs enter Earth's atmosphere they collide with a range of atoms and molecules and produce a torrent of products, including x-rays, protons, muons, electrons and neutrons, which hurtle towards Earth. Exposure to GCRs and these by-products increase with altitude. Flight attendants, residents in high altitude villages, and astronauts are all exposed to more cosmic rays than those who reside at sea-level all their lives.

Curiously, during the solar cycle, changes in the production of the solar wind modulates the fraction of the lower-energy GCR particles in a way that prevents a majority of these GCRs from penetrating the atmosphere near solar maximum. However, near the solar minimum (when sunspot activity and the formation of solar flares are minimal), GCR particles more easily penetrate the atmosphere and reach Earth's surface. Thus, the maximum rate of GCRs reaching the surface of Earth corresponds to the solar minimum.

The solar wind is very hot and travels at immense speeds, but strangely, there are two solar winds. The fast solar wind originates from coronal holes (temporary regions of cool and less dense plasma in the sun's corona). It travels at about 750 km s^{-1} and is about 800,000 °C. The slow solar wind originates from a region around the Sun's equator and has a speed of about 400 km s^{-1} but is hotter (> 1,000,000 °C). Human biology, indeed, all of biology, is unable to withstand interactions with the high energy particles and radiation of solar winds. Cancer and radiation poisoning occur when biology meets solar winds. Even the (relatively) protected electronics of satellites and space ships can be destroyed after being caught in a particularly bad period of solar flare activity. Solar panels of satellites are especially vulnerable because they must, by definition, be exposed and face the sun.

Fortunately for life-on-Earth, the magnetosphere effectively protects the surface of Earth from the worst of the solar wind.

In the absence of solar wind(s), the magnetosphere would be approximately N-S symmetrical and E-W symmetrical, like the magnetic field around a simple bar magnet (Fig. 5).

Figure 5: The symmetrical magnetic field around Earth in the absence of a solar wind. From: https://history.nasa.gov/EP-177/ch3-4.html

When a solar wind, travelling at huge speeds and containing large numbers of protons and electrons, (that is, charged particles), meets the magnetosphere, the magnetic field of the ions (when charges move, they generate their own magnetic field) and the magnetic field of the magnetosphere interact and a repulsion occurs. In addition, the solar wind, upon meeting the magnetosphere, induces an electrical field in the lower magnetosphere, although how this occurs remains poorly understood. Lorenz forces (the combination of the magnetic and electrical forces, in this case, within the magnetosphere) act upon the solar wind and deflect it around Earth. This deflection is **very important** because it protects life-on-Earth from the negative effects solar winds exert on biology. Extended periods of cells being bombarded with solar winds results in severe damage to cell structures, especially DNA (essentially, radiation poisoning) and cell function (especially protein function).

Where the solar wind approaches the magnetosphere, the velocity of the solar wind is greatly reduced and a bow shock is produced. The bow shock above Earth on the illuminated side of Earth is about 17 km thick and about 90,000 km away from Earth. The solar wind compresses the magnetosphere on the illuminated side of Earth, but on the nocturnal side of Earth, a magnetotail is created that extends much further than that seen on the illuminated side.

I should also mention the Van Allen radiation belt, which is a zone of charged particles derived primarily from the solar wind, that are captured and held by Earth's magnetic field. Earth has two Van Allen radiation belts located about 640 to 58,000 km above the surface of Earth. By trapping some of the solar wind, the magnetic field deflects those energetic particles and protects the atmosphere from being stripped by the solar wind. These belts are dangerous for satellites, which must have their electronics protected with shielding.

In summary, the Earth's magnetic field is vital to protecting life-on-Earth from the immensely dangerous solar winds that stream past Earth every second of every day.

It turns out, however, that the magnetic fields and associated magnetic poles are far from static and exhibit some very curious behaviours.......

MOVEMENT OF THE NORTH MAGNETIC POLE

Earth's magnetic field is not constant: the strength of the field and the location of its poles vary through time. The strength of the field typically ranges between 25 – 65 µT (micro Tesla) or 0.25 to 0.65 Gauss. In comparison, the field around a typical bar magnet at home is about 0.01 T (that is, 10,000 µT) or 100 Gauss. The N magnetic pole wanders a great deal (the S magnetic pole less so). We care about the movement of the magnetic poles because NATO, the US Department of Defense, the UK Ministry of Defence and the Federal Aviation Administration, as well as Google Maps and navigation systems of civilian navies and smartphones rely on knowing the precise location of the N magnetic pole. Every 5 years (or more frequently if the pole is moving quickly) the World Magnetic Model is updated to allow all of these systems to be recalibrated.

The location of the magnetic N pole (currently in the Canadian Arctic) was first established using *in-situ* measurements in 1831. For most of the 19th and 20th centuries it has been meandering slowly ($1 – 15$ km y^{-1}) towards Siberia (Fig. 6), but between 1990 and 2005 it accelerated to about $50 – 60$ km y^{-1}. In late October 2017 the magnetic N pole crossed the international date line, moving within 390 km of the geographic N pole. It is now moving southwards. The latest survey, in 2007, found the magnetic pole at 83°57'00"N 120°43'12"W. During the 20th century it moved about 1,100 km. The north and south magnetic poles also swing in an oval of about 80 km in diameter daily due to the solar wind deflecting the magnetic field.

What makes the magnetic north pole wander around so much (for some reason, the location of the south magnetic pole appears to move far less, although it is not entirely fixed). Philip Livermore, Christopher Finlay, and Matthew Bayliff, in their recent paper in Nature Geoscience (2020), demonstrated that over the past two decades the location of the position of the magnetic N pole has been mostly determined by two large-scale lobes of negative magnetic flux at the core-mantle-boundary under Canada and Siberia. What is negative magnetic flux? Hmmm. Earth's magnetic north pole actually represents the south pole of the "bar magnet residing through the centre of the Earth". The compass point labelled north points towards the Earth's magnetic north pole because the compass point is indeed the north pole of a magnet and therefore is attracted to, aligns with, points towards, what we

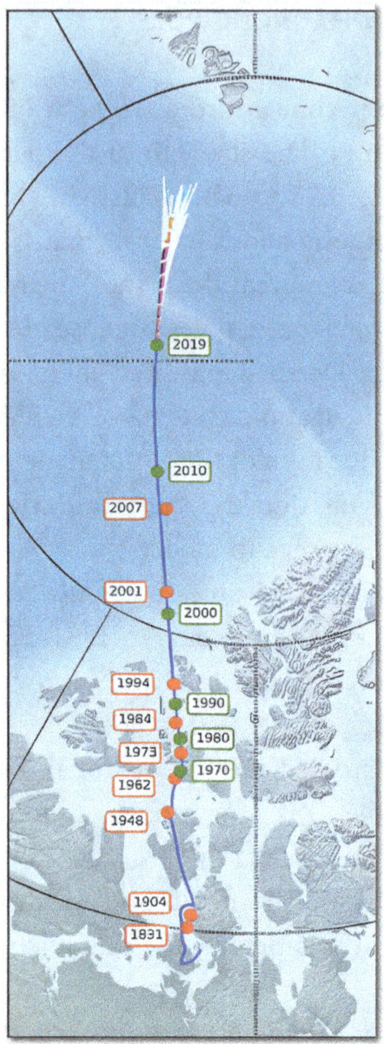

Figure 6: The trajectory of the magnetic N pole for the period 1831 to 2019. Reproduced with permission of Professor Livermore. From Livermore *et al.* Nature Geoscience 13, 387–391 (2020).

know as the magnetic north pole. This is because the Earth's magnetic north pole is really the south pole of the bar magnet within the Earth's core. Now, the magnetic field around a bar magnet is invariably drawn with an arrow directed from north to south (Fig. 7). By convention, this is deemed the positive. Outward flux (out of the N pole of the bar magnet; out of the Earth's magnetic S pole) is deemed to be positive. Inward flux is deemed to be negative. So two large lobes of the core have been identified within the northern hemisphere of the Earth's core and these have a negative magnetic field and are therefore receiving magnetic flux field lines towards them (as indeed, the entire magnetic north pole of Earth does).

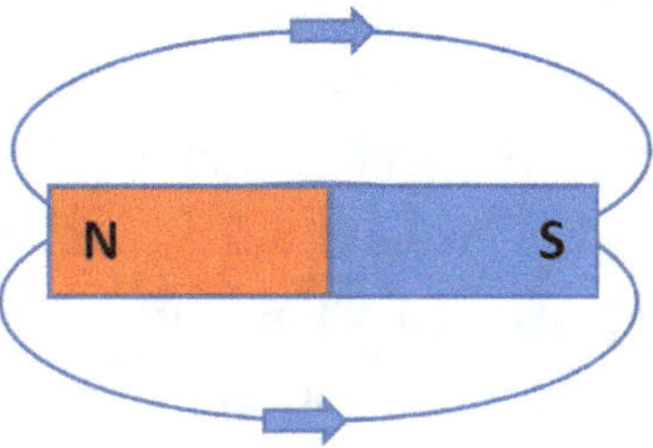

Figure 7: Magnetic field lines showing the direction of the field for a bar magnet. In Figure 5 we see the flux lines going into the Earth's magnetic N pole, telling us that the N pole is in fact the S pole of the bar magnet within Earth's core.

The north magnetic pole has travelled along an almost linear path guided along a trough of low horizontal magnetic field. This trough connects two patches of strong magnetic field, one in Canada and one in Siberia. Each of these Earth-surface patches of strong magnetic field define a magnetic dip pole (no, that isn't a typo. A dip pole is a place where the magnetic field is perpendicular to the Earth's surface) close to its centre and these two-patches of high latitude geomagnetic field define two ends of a linear channel of almost vertical magnetic field along which the north magnetic pole moves and will probably continue to move for many years into the future. So, what is the answer to the question: what makes the N magnetic pole move? The location of the N geomagnetic pole is mostly determined by the balance of a tug-of-war between the competing influences of the magnetic fields of the Canadian and Siberian lobes on the core-mantle boundary.

In conclusion, the following can be said:

Contrary to popular belief, there are 3 poles located in each hemisphere of our Earth, but the ones we are most concerned about, from a Human perspective, are the magnetic north/south poles. This is because the magnetic field surrounding Earth is really very important in protecting the Earth's surface from solar winds, a massive stream of charged particles released from the Sun's Corona; it also protects us from Galactic (and solar) cosmic rays. When biology meets solar winds, biology loses.

The location and strength of the Earth's two magnetic poles are not fixed; the strength of the magnetic field and the location of the poles vary continually. In extreme cases, the magnetic north pole can flip with the magnetic southern pole (a full flip). The last time this happened was almost 1 Mya, but the most recent partial flip occurred about 41 – 42,000 years ago. We know this from the record of magnetic field embedded in the lava that is discharged to the surface of Earth. The end of superchrons (when magnetic poles start to flip), have been associated with mass extinctions, with a 20-million-year lag between the flip and the mass extinction. The long time-lag reflects the time taken for hot mantle plumes to reach the Earth's surface and cause massive volcanic eruptions and flows of lava (flood basalt).

Large reductions in the magnitude of the magnetic field have serious implications for life on Earth because of increased levels of solar and cosmic radiation reaching the Earth's surface. Increased ionisation and decreased stratospheric ozone concentrations do bad things to life on Earth because macromolecules (DNA, RNA, proteins) are damaged by UV, ionising radiation, and solar winds. Mass extinctions of large animal species appear to be correlated with periods of large declines in the Earth's magnetic field and small differences in genetic makeup between Neanderthals and modern humans may have led to the extinction of Neanderthals in response to these changes in magnetic field.

THE END

Viruses! We have heard more about viruses and inoculations in the past 36 months than we ever wanted. COVID-19, SARS (severe acute respiratory syndrome), Spanish flu, Ebola, Monkeypox, to name just a few that have been in the news. We know many seem to have a dislike for Humans. But what are viruses, and are they alive? How do we define 'living'? How do viruses infect us? Where did COVID-19 come from? A truly rich tapestry of themes to be discussed in this chapter.

WHAT ARE VIRUSES AND WHAT ARE THEY MADE OF?

A good definition of a virus is as follows:

> "A virus is a submicroscopic infectious unit that can only replicate inside the *living cells* of an organism".

The living cell (known as the host cell) might be part of a multicellular animal (e.g., Humans, dogs, fish), a multicellular plant (e.g., a crop species, a tree), a multicellular fungus, or a single-celled organism such as a bacterium or an amoeba (for example). The virus itself is generally not considered to be a cell in its own right because it lacks almost all of the attributes of a living cell (more on this later). They are also very small. Most animal and plant cells are in the range 0.01 to 0.1 mm in diameter, or 10 – 100 µm. Human hair is typically about 50 – 100 µm in diameter. Viruses are about 20 – 300 nm (or 0.02 – 0.3 µm) in diameter. A *really* big virus can be up to 500 nm (0.5 µm). There are 1000 nm in 1 µm, so a virus with a diameter of 100 nm has a diameter of 0.1 µm. To put this into a visual perspective:

If this line (above) represents the 100 μm diameter of a large plant cell, then the following line represents the diameter of a **large** virus (not so much a line as a dot):

.

So what are viruses made of?

All viruses have two components, but some viruses have three components. The two components common to all viruses are: (1) a protein shell (called the capsid); this surrounds (2) the genetic material (or genome) of the virus, that is, either DNA (deoxyribonucleic acid) or RNA (ribonucleic acid). Some viruses have a third component, a lipid membrane (and associated proteins) around the protein shell. This lipid membrane (and associated proteins) is derived from (pinched off from) the host cell when the virus erupts from an infected host cell.

THE CAPSID

The protein shell is not a single protein but consists of many copies of two or more proteins. These two (or more) proteins come together to form a protomer, and then multiple copies of these protomers aggregate to form a capsomere. Many copies of the capsomere self-aggregate to come together to form the capsid (the complete coating around the nucleic acid of the genome). So the protein shell of a virus is made of multiple copies of a capsomere and each capsomere is made of two or more proteins, arranged together to form a protomer. The final shape of the capsid can be used to provide a classification of viruses and the five shapes commonly seen are helical, icosahedral (which are almost round, but having 20 faces, a bit like a soccer ball), prolate (like a rugby ball), enveloped (not a shape at all, but a useful classification for viruses), and complex (also called head and tail; see Fig. 1).

Helical viruses are made from multiple copies of a single type of capsomere, all stacked around a central axis to form the helix. Such viruses might be short, rigid rods, or long flexible filaments. The genetic material (often single-stranded RNA, but occasionally single-stranded DNA) is attached to the protein helix.

Icosahedral or near-spherical viruses are the most common type of virus infecting animal cells. A regular icosahedron is the optimum way of forming a closed shell from identical subunits (like a soccer ball is made up of hexagonal and pentagonal panels).

Prolate viruses are composed of an icosahedron elongated or 'stretched' along an axis, with two endcaps attached.

Enveloped viruses are wrapped in a modified form of host cell membranes. This membrane has many different proteins attached/embedded, some of which are coded by the genome of the virus, and others are coded for by the genome of the host cell. Influenza virus, HIV (human immunodeficiency virus), and severe acute respiratory syndrome coronavirus 2 (which causes COVID-19), are enveloped viruses and it is the envelope that confers their infectivity.

Complex viruses look like aliens (Fig. 1). They typically possess extra structures, including a protein tail, and are called bacteriophages. The tail acts like a molecular syringe, attaching to a soon-to-be infected bacterial host cell and then injecting the viral genome into the cell.

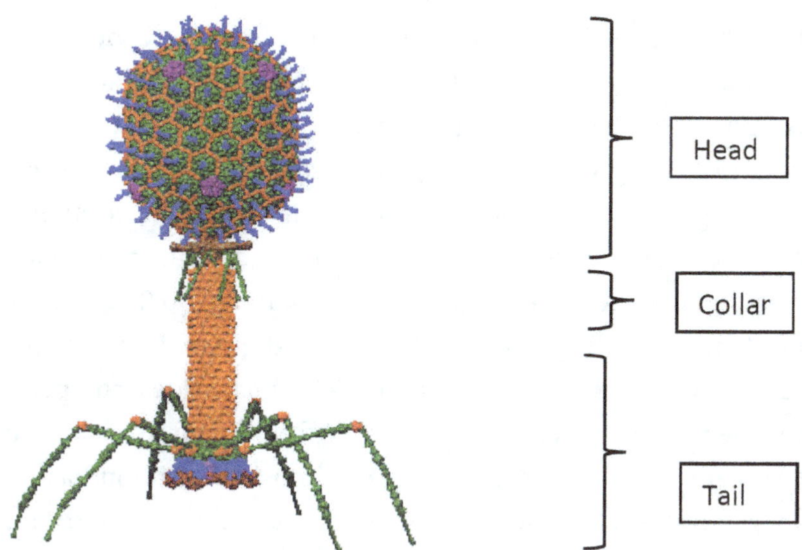

Figure 1: *Escherichia virus T4* is a bacteriophage that infects *Escherichia coli* bacteria. It is a double-stranded DNA virus. Drawn by Dr. Victor Padilla-Sanchez, PhD. This image is licensed under the Creative Commons Attribution-Share Alike 4.0 International license. From https://en.wikipedia.org/wiki/Bacteriophage.

DNA, RNA, AND *THE GENETIC COMPOSITION OF VIRUSES*

Chromosomes in the nucleus of a human cell consist of a very long strand of DNA, with as many as 500 million base pairs (see text box for the definition of a base pair) making up thousands of genes on each chromosome. Each gene must be transcribed (copied) into an equivalent RNA molecule to get the information (the genetic code of the gene) out of the nucleus and into the cytoplasm, where the metabolic machinery of the cell is located. If the gene on the chromosome encodes the synthesis of a protein, a single stranded RNA is produced called 'messenger RNA' (mRNA) and this mRNA corresponds to the nucleotide sequence of the gene on the DNA. The message contained in the mRNA is used to direct the synthesis of a specific protein.

The genetic material in viruses can be DNA (the same as the DNA in the nucleus of human cells) or RNA (the same as the RNA that is found both in the nucleus and outside of the nucleus of human cells)). The genetic material of viruses can be single stranded (the same as mRNA in human cells) or double stranded (the same as DNA in the nucleus of human cells).

What does this viral RNA/DNA code for? The phrase "code for" means "the information required to synthesise a protein". The genome (total genetic material) of a virus is very, very, small and tightly packed inside the capsid. Viruses manage to infect a host cell and reproduce millions of copies of themselves using this remarkably small genome. For example, the flu virus genome contains only about 14,000 bases on 8 single-stranded RNA segments, and contains 8 genes and codes for 11 proteins, while the human genome contains 3,200,000,000 bases (about 200,000 times longer), coding for more than 20,000 genes. Bacteriophage Qβ, one of the smallest known RNA viruses, contains 4217 nucleotides, comprising 4 genes. In contrast, some viruses are veritable giants (within the viral world). For example, the giant *Megavirus chilensis* has a genome of almost 1.3 million bases, coding for perhaps 1000 genes. I briefly discuss what genes are and their importance after the text box on base pairs.

What is a base pair?

A base pair is the basic building block of double-stranded DNA found in the nucleus of human cells (and the cells of millions of other species of animals, fungi, and plants). A base pair consists of two nucleobases (or simply, bases) bound to each other by hydrogen bonds. A base is a 5 or 6 six-sided ring containing C and N (Fig. 2). One way of thinking of double-stranded DNA is to think about a ladder that has been twisted. The two vertical sides of the ladder represent the bases of the two intertwined helices and the rungs of the ladder represent the hydrogen bond linking the two helices.

One of the bases of a base pair is part of one of the two helices of double-stranded DNA and the other nucleobase is part of the other helix of double-stranded DNA. There are four bases in human DNA: guanine (G), cytosine (C), adenine (A) and thymine (T). In RNA, the base uracil (U) replaces thymine. G always binds with C and A always binds with T (or U, in RNA). The A-T base pair (below) is linked via 2 hydrogen or 'H' bonds (the dotted lines in the figure below). The G-C pair is linked by 3 H-bonds. The human nucleus contains 23 chromosomes, with a total length of about 3.2 billion bases, coding for 20,000 – 25,000 distinct protein-coding genes.

Each of the 4 bases is attached to a sugar molecule and a phosphate group. A base plus a sugar molecule plus a phosphate group makes a nucleotide. The nucleotides are joined together via covalent bonds to form a single strand of DNA, and 2 strands are joined by H-bonds to form the classical helix structure of double stranded DNA.

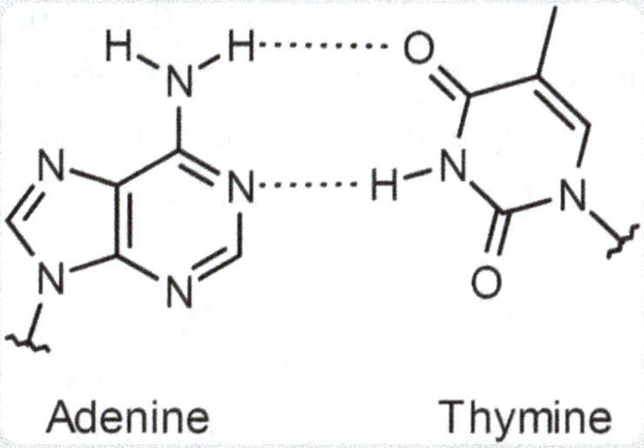

Figure 2: A base pair consisting of adenine and thymine, linked by hydrogen bonds (the dots between the H and O and H and N of the two bases).

With so few genes present, how does a virus manage to fulfill its destiny (that is, infect a host cell, replicate, and emerge to infect new cells/hosts)? There are two types of genes in viruses, bacteria, plants, animals and fungi. The first contain the information required to synthesise RNA, which is a large molecule that is used as a template for the synthesis of proteins. Proteins are what enzymes are made of. Enzymes are responsible for catalysing specific chemical reactions within cells. Other proteins are required to move ions (for example, K^+) and molecules (for example, sugars) across membranes. Without these proteins, life would not be possible. The other type of gene are called non-coding genes. Once thought of as "junk" DNA, we now know that many such non-coding genes have important roles, including the control of gene activity. So the question "how does a virus, with so few genes, fulfill its destiny is an interesting one and I address it shortly, but first, the question: are viruses alive?

Are viruses alive?

A good question. What does "being alive" mean? This is a vexed question, but living organisms generally, I believe, have several common features, as follows:

1. Living organisms are cellular and perhaps most importantly, the cells contain an array of sub-cellular structures (e.g., a nucleus, Golgi apparatus, mitochondria, in eukaryotes, or *identifiable compartments* in prokaryotes and even some membrane-bound organelles in some prokaryotes) that perform all the functions required to keep the cell alive. However, not all cells contain all the sub-cellular structures found in living cells. For example, red blood cells do not contain a nucleus, which presumably is why they don't live for very long (compared to other cells of the human body). Similarly, the epidermal cells of leaves don't contain chloroplasts and so epidermal cells don't photosynthesise.

2. Life requires consumption of energy. At a cellular level, the energy currency of cells is a molecule called adenosine tri-phosphate (ATP). All living cells (bacterial, fungal, plant, and animal cells) contain ATP. The functions of ATP are extremely varied, including: a source of stored energy to drive

muscle contraction; powering ion movement across membranes; activation and inhibition of many enzymes; DNA and RNA synthesis; a neurotransmitter in many parts of the nervous system. The energy consumed by human cells in an adult requires the hydrolysis (to release the energy stored in the molecule) of 10 billion molecules of ATP every day. ATP is very effectively recycled, and each gram of ATP is used very many times each day (more than 1000 times per day) (See ATP textbox).

3. The cells of living organisms reproduce themselves using a large array of sub-cellular *structures* or compartments (for example, ribosomes (the site of protein synthesis), endoplasmic reticulum (an interconnected network of flattened, membrane-enclosed sacs), and actin (a protein in the sub-cellular scaffolding that can move organelles like mitochondria and chloroplasts around) (or its homologue in prokaryotes)) and *processes* to replicate proteins, DNA, RNA, lipids, the nucleus, mitochondria, and much else besides.

4. Living organisms evolve across generations.

5. Living cells exhibit homeostasis, whereby they regulate the internal pH, salt and water content, and metabolism, such that intracellular conditions are as close to optimal as can be achieved.

So, do viruses share these features of living organisms? Well, they are not cellular, they do not, of themselves, contain, synthesise or consume ATP, and, in the absence of a host, they can't replicate. They don't exhibit homeostasis because they lack any metabolism within their protein coat. But, it must be said, they do evolve, rapidly and continually.

So I conclude, as have many others, that they are not alive. But this does beg the question……. What is life? An amoeba and a bacterium are bags of chemicals undergoing a sequence of regulated reactions, including synthesis and breakdown of many types of molecules, energy conversions, and homeostasis, and they evolve. But where is the line drawn, between a very small stripped-down bag containing a few chemicals (a virus), and a larger, more complex bag of chemicals (e.g., a bacterium, an amoeba, a human cell)?

Adenosine Triphosphate (ATP)

Physicists and engineers tell us that work is done when a force is applied to a mass and the mass moves in response to that force. For example, when a car (the mass) is driven (moves) along a road, petrol (or diesel or electricity) is the energy supply used to generate a force, via an engine (or electric motor) that is transmitted to the wheels via a gearbox and drive shaft. Similarly, when we lift a mug of tea to our lips, our muscles are doing work to lift the mug of tea. What is the energy supply used in muscles?

At a cellular- and sub-cellular-scale, work is done when muscle fibres contract, when ions (for example Cl^- or NO_3^-) are pumped into cells across membranes, when the flagellum of mammalian spermatozoa beat, and when the numerous cilia of ciliated protozoa (for example, *Paramecium*) beat to propel it through water. Within cells there is a large array of molecular motors which perform mechanical work; they are essential for the movement of proteins, vesicles and sub-cellular organelles (for example, mitochondria and chloroplasts) and the movement of chromosomes during cell division.

The universal energy supply for all these work-related processes is the molecule adenosine triphosphate (ATP). My biochemistry lecturer at the University of Sussex in England used to tell us that the synthesis and degradation of ATP was the purpose of life. Of course, she was wrong: the synthesis and degradation of ATP is the *definition* of life, not the *purpose*. Life has no purpose.

ATP is composed of adenine (one of the four nucleobases of DNA) attached to a 5-C sugar (ribose) which is attached to three phosphate groups (see below).

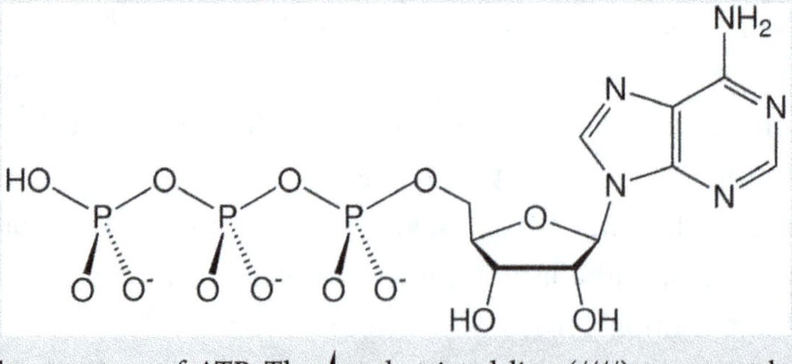

Figure 3: The structure of ATP. The ⬆ and striated line (/////) represent bonds that are directed out of the page towards us and into the page, respectively.

The key point, however, is that the P-O-P bonds (there are two such bonds) are high energy bonds. When ATP is hydrolysed to ADP (representing the loss of a single inorganic phosphate group, P_i) the free energy released is 30.5 kJ mol^{-1}. If both P-O-P bonds are hydrolysed (thereby releasing PP_i) 45.6 kJ mol^{-1} of free energy is released. It is this released energy that is used by molecular motors, molecular pumps, and cilia and flagella to perform work, that is, to move stuff into, out of, and within, cells.

The question: are viruses alive? has become even more complicated over the past two decades with the discovery of giant viruses in amoebae. The one known species of giant virus (*Megavirus chilensis*) contains more genes than a simple bacterium and contain genes for information storage and processing. The *Megavirus chilensis* genome is composed of a linear, double-stranded molecule of DNA (Humans have double-stranded DNA), and is 1,259,197 base pairs in length, making it the second largest viral genome on record. The Megavirus genome contains more than 1100 protein-coding genes, a number larger than most bacteria. The largest genus of viruses is the Pandoravirus, discovered in 2013. These viruses have double-stranded DNA genomes, possess the largest genome (2.5 million base pairs) of any known viral genus, and infect amoeba. Weirdly, genes common to the three known life domains (Archaea; single-celled organisms lacking a membrane-bound nucleus), Bacteria (single-celled organisms lacking a membrane-bound nucleus, differing from Archaea in their genetic composition, metabolism, and membrane composition), and Eukarya (organisms with a membrane-bound nucleus, including algae, crops, trees, dogs, insects, and Humans) can be found in different giant viruses. Consequently, some researchers argue that the giant viruses constitute a *fourth domain of life* and are therefore alive. Some virologists argue that viruses may have originated from early RNA-containing cells, but evolved away from the cellular life form, to become a hyper stream-lined existence. *Did they, in fact, evolve out-of-life, and back into pure chemistry?*

A final complicating factor in considering the question: what is life? Prions and plasmids are two recently discovered structures consisting of proteins (prions) or DNA (plasmids). Prions are misfolded proteins which induce normal variants of the same protein to become misshapen. They cause several fatal, and *transmissible,* neurodegenerative diseases in Humans. Kuru, for example, is a very rare and fatal neurodegenerative disease that was, at one time, relatively common among the Fore people of Papua New Guinea. Kuru is a transmissible spongiform encephalopathy caused by the transmission of prions, that is, proteins that have not folded properly into the usual 3-D conformation. It is transmitted through eating contaminated Human brains, a practice once common in funerary cannibalism (where members of the family of the deceased eat some of the body of the deceased). The word prion derives from "proteinaceous infectious particle" and the idea of a protein being an infectious agent is in stark contrast to viruses, bacteria, fungi, and parasites because, by definition, a prion does not contain any nucleic acids (RNA, DNA). In marked contrast, plasmids are very small, extrachromosomal, double-stranded DNA molecules within a cell; they are physically separated from chromosomal (that is, nuclear) DNA and can replicate independently of nuclear DNA. They are often circular, occasionally linear, and occur in bacteria and *Archaea* and eukaryotic organisms, carrying genes that benefit the survival of the host organism, including conferring antibiotic resistance to bacteria, or the ability to fix atmospheric nitrogen. Plasmids are replicons, units of DNA capable of replicating autonomously within their host cell. Plasmids are transmitted from one bacterium to another (even of another species) mostly through conjugation, and this constitutes horizontal gene transfer (that is, in contrast to vertical gene transfer from "parent" to "offspring" when a bacterial cell divides into two cells). Plasmids are "naked" DNA of 1 to over 200 thousand base pairs. The number of identical plasmids in a single cell can range from one to several thousands. Artificially-constructed plasmids are used as vectors (carriers) in genetic engineering, to clone and amplify (make many copies of) specific genes. They are also used to induce bacteria to makes lots of useful protein, for example, insulin, for the treatment of diabetes.

It would be hard to consider either prions (a bit of protein) or plasmids (a bit of DNA) as alive – wouldn't it?

So it all comes down to which definition of life you ascribe to. I like my five attributes of life definition (see above), but an alternative is that life is something that is capable of genetic evolution, which viruses certainly are. However, the centrality of ATP, plus homeostasis and genetic evolution, for me, are key to defining life.

HOW DO VIRUSES INFECT CELLS AND REPLICATE

To infect a host cell, an individual virus particle (called a virion) must enter the cell. There are two ways to do this. But there is one vital step prior to entering the host cell. First, the virion must attach to the host cell. In animal cells the host cell is surrounded by a plasmalemma (also called a plasma membrane). This is what keeps the bag of chemicals just that – a bag of chemicals (Fig. 4).

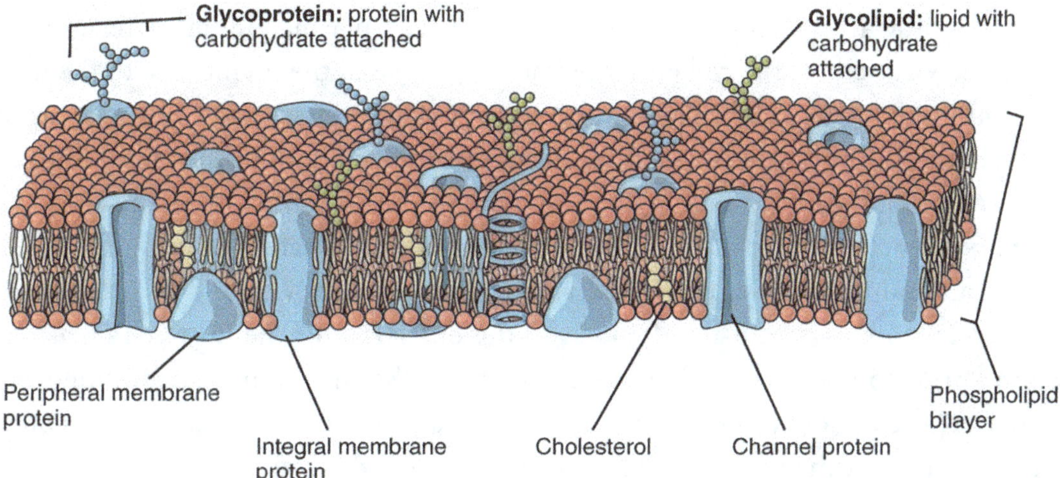

Figure 4. Cell membranes are composed of a phospholipid bilayer within which proteins are embedded. These proteins may be visible only on one side (peripheral membrane proteins), or they may traverse the whole width of the membrane (integral membrane proteins). These proteins may form channels for the movement of water, or they may be energy-consuming pumps that move ions across the membrane. From wikipedia https://en.wikipedia.org/wiki/Biological_membrane URL: https://cnx.org/contents/FPtK1zmh@8.108:q2X995E3@12/The-Cell-Membrane. This work has been released into the public domain by its author, LadyofHats (Mariana Ruiz).

The plasmalemma is made of a phospholipid bilayer (phospholipids are a class of lipids having a *hydrophilic* (likes being in contact with water) head containing a phosphate group, and two *hydrophobic* (doesn't like being in contact with water) tails derived from fatty acids, joined by a glycerol molecule. The hydrophilic heads of the two layers face outwards and therefore interact with water molecules. The hydrophobic tails of each layer face inwards and don't interact with water. Several proteins stick out of the plasmalemma on both sides (Fig. 4) of the membrane;

these are called receptor proteins, and their function is to attach to a wide range of chemicals and then signal this fact to the interior of the cell. This is what viruses have evolved to manipulate. But the method of entry to a host cell differs according to whether the virus is an enveloped virus (having a membrane around the protein shell; the flu virus is an enveloped virus), or whether it is a naked virus (lacking a membrane around the protein shell; the poliovirus is a naked virus).

A naked virus contains specific proteins as part of the capsid (the protein shell surrounding the DNA/RNA of the virus) which attach to a receptor protein (also called an attachment factor) located on a host cell plasmalemma. After attachment, the virus enters the cell either by making a hole in the membrane of the host cell and inserting its viral genome, or by endocytosis. Endocytosis is a process of inward budding of the plasmalemma (invagination) to surround the virus. A small membrane-bound bag containing a virus particle moves into the cell and then the membrane surrounding the virus is split open and the virus released into the cell. Poliovirus and Hepatitis-C are naked viruses that enter cells by endocytosis.

Enveloped viruses must still attach to the outer face of the host cell's plasmalemma prior to entering the host cell. This time, the membrane surrounding the viral protein coat contains the proteins that interact with the receptor proteins poking out of the plasmalemma of the host cell. Enveloped viruses also contain fusion proteins as part of the membrane surrounding the protein shell. These fusion proteins cause the membrane envelope of the virus to fuse with the plasmalemma of the host cell. These fusion proteins are derived from the virus' genetic information. Because the composition and structure of a fusion protein are unique to each type of virus, broad spectrum antiviral drugs have yet to be made, and specific vaccines have yet to be developed for each type of enveloped virus. Those vaccines that have been developed against enveloped viruses are targeted so that the Human body produces antibodies against the envelope fusion proteins.

Following fusion of the membranes, a pore is formed through the action of the fusion protein of the virus, and this pore becomes filled with water and the genetic material of the virus diffuses out of the virus and into the interior of the host cell. Alternatively, some enveloped viruses are taken up by endocytosis, as occurs for some naked viruses.

BACTERIOPHAGES

There is one additional method of getting the genetic material of a virus into a cell, a method employed by bacteriophages, the so-called head-and-tail viruses (Fig. 1). Bacteriophages infect bacteria and *Aracheea*.

Bacteria possess a cell wall made of polysaccharides. To infect a bacterial host cell, bacteriophages bind to specific receptors on the surface of bacteria. To get through the polysaccharide coating, enzymes on the surface of the virus break down the polysaccharide cell wall.

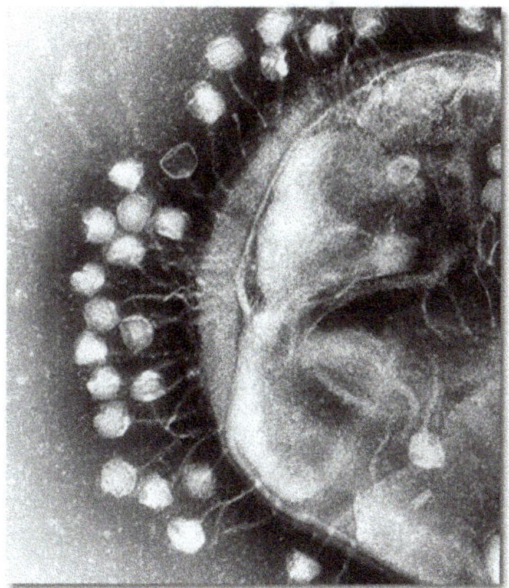

Figure 5:   An electron micrograph of bacteriophages (the white square(ish)/hexagonal-like heads with long thin tails) attached to a bacterial cell. The viruses are the size and shape of coliphage T1. This image is licensed under the Creative Commons Attribution-Share Alike 3.0 Unported license. From https://en.wikipedia.org/wiki/Bacteriophage#/media/File:Phage.jpg

Bacteriophages use a structure resembling a hypodermic syringe (Figs. 1, 6) to inject their genetic material into the cell. Shortly after attaching to the bacterial cell at a receptor protein, the tail fibres flex to move the base plate towards the surface of the bacterial cell. The tail then contracts and injects the viral genetic material through the bacterial membrane.

The virus is now inside the cell, but the genetic material (DNA or RNA) of the virus (naked or enveloped) remains inside the virus (with the exception of bacteriophages which inject the genetic material into the cell). Now what?

Four steps remain before the virus has completed its "life-cycle". These are:

1. uncoating, where the viral capsid (protein coat) is removed and degraded by enzymes, thereby releasing the viral nucleic acid;
2. replication, where the viral genome is transcribed or translated (see below). At the end of replication, new copies of the viral proteins and genetic material will have been produced, allowing the next step to start;
3. assembly, where viral proteins are assembled into new viral particles, each containing a copy of the viral genome. This triggers the last stage;
4. release from the host cell. Two methods of viral release are recognised: lysis or budding. Lysis kills infected host cells, and smallpox is one such virus using this method. Enveloped viruses, (e.g., the influenza A virus, COVID-19) are usually released from the host cell by budding, and this is what gives these viruses the membrane coating, entirely derived from the host plasmalemma. Host cells are usually not killed by this process.

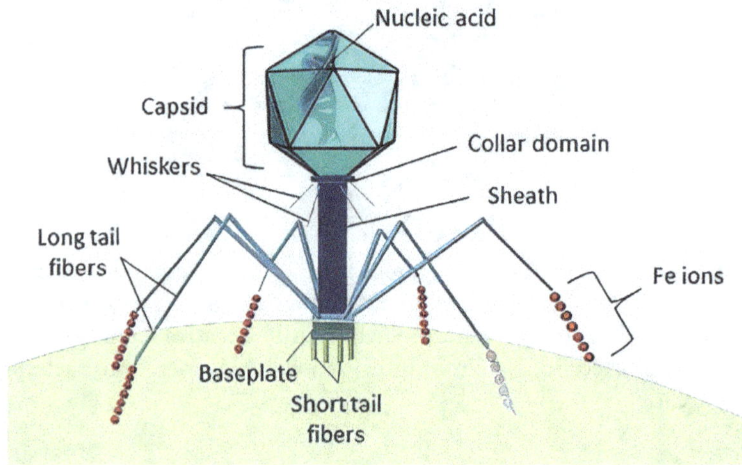

Figure 6: A schematic of a myovirus bacteriophage showing the head (capsid) containing the nucleic acid, and the syringe-like tail used to inject nucleic acid into the bacterial cell. This image is licensed under the Creative Commons Attribution-Share Alike 4.0 International license. Authors are Chelsea Bonnain, Mya Breitbart and Kristen N. Buck Image from https://en.wikipedia.org/wiki/Myoviridae#/media/File:Structure_of_a_Myoviridae_bacteriophage_2.jpgwiki

The DNA or RNA contained within a virus is released after the breakdown of the viral protein coat (and membrane, if a coated virus) within the host cell.

RNA viruses contain RNA and this RNA contains the instructions required for the synthesis of proteins. A suite of proteins must be synthesised by the pirated host-cell, including structural proteins for the capsid, fusion proteins, transmembrane proteins, and enzymes for RNA replication.

When there are more than one RNA viral genomes present in a single cell, the RNA of the two genomes are able to combine, mix their genetic composition, and give rise to new, mutated, RNA viruses, a process called recombination.

Viruses containing DNA, not RNA, contain either single-stranded or double-stranded DNA. DNA viruses that cause disease in Humans include herpesviruses, papillomaviruses, and poxviruses. Generally, the DNA must enter the nucleus of the host cell. This is achieved by active movement through the nuclear pore in the nucleus envelope, or through DNA injection through the nuclear membrane. Once in the nucleus, the host cell's machinery is hijacked to replicate the viral DNA.

The next step is the assembly of new viral particles from newly synthesised RNA/DNA and proteins (capsid proteins, plus membrane bound proteins if the virus is an enveloped virus). Assembly is primarily driven by hydrophobic interactions (the interaction of water and hydrophobic (i.e., water repelling) molecules, electrostatics (attraction or repulsion of charged atoms/molecules), van de Waals forces (the attraction of charges on atoms or molecules; it is a very weak, yet very important, type of intermolecular interaction), and hydrogen bonding (an electrostatic force of attraction between a hydrogen atom and an electronegative atom with a lone pair of electrons). Inclusion of the genome within the protein shell (the capsid) arises in many instances because of electrostatic interactions between positive charges on capsid proteins and negative charges on RNA/DNA. Thus, viral assembly is driven by physics (or chemistry, depending on your philosophical outlook) and not by the internal "biochemical machinery" of the host cell.

The final step is the release of the virus from the host cell. This can be achieved through rupture of the host cell, which kills the cell and releases the viruses into, for example, the blood stream. Alternatively, a virus can be released by budding, which is how enveloped viruses get their lipid membrane. The assembled virus is surrounded by a portion of the outer membrane of the host cell, which protrudes outwards and buds off, releasing the virus, with the associated host-cell membrane and glycoproteins within the membrane. Binding of the capsid protein to membranes occurs because of electrostatic interactions between positive charges on capsid proteins and negative charges of lipid heads. Thus, viral release, just like viral assembly, is driven by physics (or chemistry).

COVID-19: A QUICK REVIEW

SARS-CoV-2 is the virus that caused the recent pandemic, of 2020, 2021, and 2022. What do we know about it? The disease it causes is named COVID-19, which arises from the words Corona (because it sort-of looks like a crown when viewed with an electron microscope; Fig. 7), virus, disease, and the year 2019, when it was first identified, in Wuhan, China. COVID-19 is a contagious disease and is responsible for approximately 7 million deaths world-wide, so far. The most commonly observed symptoms include, a cough, loss of smell and taste, headache, nasal congestion and runny nose, muscle pain, sore throat, fever, and breathing problems.

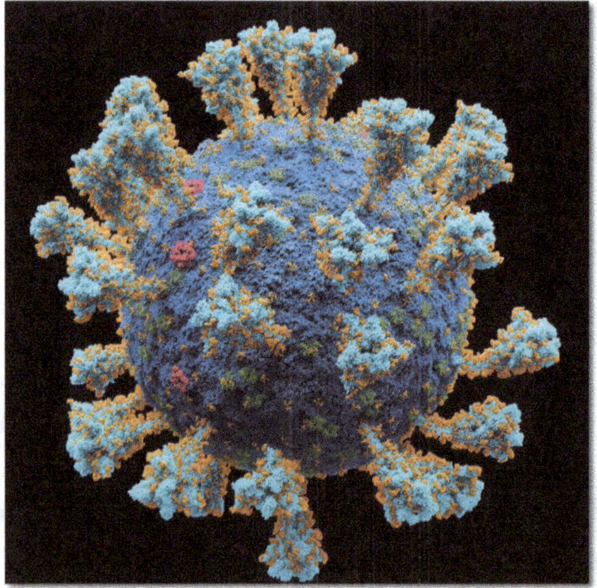

Figure 7: A representation of the external features of the SARS-CoV-2. Each differently coloured ball represents a different atom. Authors: Alexey Solodovnikov and Valeria Arkhipova and numerous consultants. Licensed for free use for commercial and non-commercial uses. From https://en.wikipedia.org/wiki/COVID-19#/media/File:Coronavirus._SARS-CoV-2.png

SARS-CoV-2 is an enveloped virus. It has a lipid membrane surrounding the protein coat that surrounds the interior, which contains single-stranded RNA, not

DNA. This single strand of RNA can be directly "read" or directly "translated" to produce a protein. The structural proteins encoded in the RNA of SARS-CoV-2 include a membrane glycoprotein, an envelope protein, the nucleocapsid protein, and the spike protein. The function of the spike protein is to allow viral entry into the host cell by interacting with receptor molecules on the outer surface of a human cell and then fusing viral and cellular membranes.

Genetic analysis has demonstrated that the coronavirus genetically clusters with the genus Betacoronavirus, and is closely related to two bat-derived strains. It is 96 % identical, at the whole genome level, to other bat coronavirus samples. The membrane glycoprotein of SARS-CoV-2 is about 98 % identical to the membrane glycoprotein of bat SARS-CoV and exhibits about 98 % homology with pangolin SARS-CoV. These similarities have been taken, by some, to mean that SARS-CoV-2 is a zoonotic virus, meaning it was originally found in animal species (bats and pangolins) but very recently jumped to humans. This is discussed in the next section.

WHERE DID SARS-CoV-2 *COME FROM?*

There are two views on the origin of COVID-19. The first view is that it has a zoonotic origin, meaning that it originated in another species (perhaps a bat or a pangolin (a scaly anteater)) but at some (recent) stage, it acquired the capability of infecting humans. This origin is not uncommon. For example, ebola, plague, rabies, and brucellosis are zoonotic diseases. Ebola and rabies are viral diseases, plague and brucellosis are bacterial. Interestingly, recent(ish) coronavirus outbreaks are also zoonotic. For example, severe acute respiratory syndrome (SARS), which appeared in 2002 in China and spread globally within a few months, and Middle East respiratory syndrome (MERS; first reported in Saudi Arabia in 2012), are both viral respiratory illnesses new to Humans.

The second view is that COVID-19 is the result of an accidental escapee from a research lab, and in particular, the Wuhan Institute of Virology, located in the city of Wuhan, China.

Understanding the origins of SARS-CoV-2 is critical for the development of future global strategies to minimise the impact of future pandemics. Let's take an objective look at evidence for both theories.

A Zoonotic origin

In May 2020, the World Health Assembly requested that the World Health Organization (WHO) Director-General work with partners to assess possible origins of SARS-CoV-2. In November, the Terms of Reference for a China–WHO joint study were released and then information, data, and samples for the study's first phase were collected and summarised by the Chinese members of the review team. Although there was clear support for both a zoonotic origin or an accidental leak from a research lab, the review concluded that a zoonotic spillover from an intermediate host was "likely-to-very-likely," and an accidental leak was "extremely unlikely". However, the two possibilities were not given equal time in the assessment or the report itself. Only 4 of the 313 pages of the report and its annexes discussed the possibility of a laboratory accident. Even the WHO Director-General Tedros Ghebreyesus noted that the report's consideration of evidence supporting a laboratory leak was insufficient. Indeed, a letter published in the prestigious journal Science, in May 2021, written by 18 scientists with relevant experience, indicated "…. greater clarity about the origins of this pandemic is necessary and feasible to achieve. We must take hypotheses about both natural and laboratory spillovers seriously until we have sufficient data." But, the zoonotic origin remains a viable hypothesis.

Evidence for the zoonotic origin theory includes the following:

1. Distinctive features of the SARS-CoV-2 genome argue against it being a lab-leak. Bats are known carriers of coronaviruses, and the genome of SARS-CoV-2 is closest to that of a coronavirus first found in a horseshoe bat in the southern Chinese province of Yunnan (where Wuhan is located) in 2013. However, this genome is only 96 % identical to the genome of SARS-CoV-2. A closer genetic relative of the virus, that is, the one passed to Humans, remains unknown.

2. It is unlikely that SARS-CoV-2 emerged through laboratory manipulation of a related SARS-CoV-like coronavirus because: (a) although the receptor binding domain (RBD) of the spike protein of SARS-CoV-2 appears to be optimised for binding to Human receptor proteins, the mechanism of

such optimisation appears to be different to previous predictions of how to optimise such binding; and (b) if genetic manipulation *in vivo* had been undertaken, it is most likely that one of the several reverse-genetic systems available for betacoronaviruses would have been used, but genetic analyses of the virus show that SARS-CoV-2 is not derived from any previously used virus backbone. Therefore, two scenarios can explain the origin of SARS-CoV-2 in a zoonotic origin scenario: (i) natural selection in an animal host before zoonotic transfer; and (ii) natural selection in Humans following zoonotic transfer.

3. Natural selection in an animal host before zoonotic transfer. Many early cases of COVID-19 were linked to the Huanan market in Wuhan, where live animals are traded (a so-called "wet market"). Consequently, it is possible that an infected animal source was present and traded in the market, including bats and pangolins. Malayan pangolins imported into Guangdong province contain coronaviruses similar to SARS-CoV-2. Although the bat virus remains the closest to SARS-CoV-2 across the genome, some pangolin coronaviruses exhibit strong similarity to SARS-CoV-2 in the receptor binding domain of the spike protein. It has been concluded by many that this demonstrates that the SARS-CoV-2 spike protein is optimised for binding to Human-like ACE2 as a result of natural selection, and not Human modification.

4. Natural selection in Humans following zoonotic transfer. It is possible that an ancestor of SARS-CoV-2 was transmitted to Humans, and then acquired the distinctive genomic features during Human-to-Human transmission. Once these adaptations evolved, they enabled the pandemic to take off. All SARS-CoV-2 genomes sequenced so far have the genomic features described above and are thus derived from a common ancestor that had them too. The presence in pangolins of an receptor binding domain very similar to that of SARS-CoV-2 means that this could also have been in the virus that jumped to Humans.

But, what of the alternative view?

An accidental lab-leak

Evidence for the *alternate hypothesis,* namely, that the virus originated in the Wuhan Institute of Virology (WIV) and accidently escaped, is in short supply. In the absence of a full, free, independent, and in-depth examination of the Institute, its scientists, facilities and records, it is unlikely that definitive evidence will be discovered to support this hypothesis (this goal was not achieved by the WHO investigation of early 2021). But is there any circumstantial evidence that might support the hypothesis?

1. The WIV was certainly engaged in bat virus research for several years prior to 2020, so the potential for a leak of a bat virus exists, but this is certainly no smoking gun because many labs around the world are engaged in similar animal virus research. However, it is consistent with the hypothesis.

2. Dr Ai Fen, the Head of Emergency Medicine at Wuhan Central Hospital, was interviewed by a Chinese magazine in 2020 about a new fever that was being increasingly detected in Wuhan. Dr Fen obtained genetic analyses of the new virus and it was identified as a novel coronavirus. That interview was erased within 24 h of it being uploaded and she reported that she had been reprimanded for "spreading rumours". Again, not a smoking gun, but worrisome.

3. A colleague of Dr Ai Fen, Dr Li Wenliang, also working at the Wuhan Hospital, informed a number of people on a WeChat group of multiple cases of SARS in late December 2019. He too, was later punished by the Chinese Government for discussing the novel fever. Again, not a smoking gun, but certainly worrisome.

4. Dr Shi Zhengli is a world renown virologist at the WIV. She has been conducting genetic research on bat viruses for many years and has manipulated spike protein genes for many years. In 2005, she and several colleagues found that bats are the natural reservoir of SARS-like coronaviruses. She then led a research team examining binding of spike proteins of both natural and chimeric (containing DNA/RNA from 2 or more different types of virus) SARS-like coronaviruses to angiotensin-converting enzyme 2 (ACE2) receptors in Human, civet, and horseshoe bat cells. ACE2 is an

enzyme located on the surface of many types of cell. It produces small proteins by cutting up the large protein angiotensinogen. The SARS-CoV-2 virus binds to ACE2 using its spike-like surface protein as a first step to infecting a cell. ACE2 is like a cellular doorway. Not a smoking gun, but indicative of the skills and types of research being undertaken at the WIV relevant to the hypothesis.

5. Chinese researchers tested about 80,000 animals, both wild and farmed, around Wuhan and did not find any evidence of SARS-CoV-2 infection. This certainly doesn't prove that it isn't, or wasn't, out there in an animal population around Wuhan, but it means if it is, it is quite rare.

6. A senior PLA scientist filed a patent application for a coronavirus vaccine in February 2020, a truly remarkable feat to combat a pandemic that had only been declared a few weeks earlier.

7. In early September 2019, the Institute's virus database disappeared from its webpage for almost a year. Certainly not a smoking gun but undoubtedly worrisome and this does beg two questions: what was being hidden by removing the webpage for so long, and why did it take so long for the webpage to be re-installed?

8. A series of high-level and presumably highly-informed individuals have categorically stated that the evidence they have seen points to a lab-leak. These people include Sir Richard Dearlove, former head of MI6, Mike Rogers, a former US Navy Admiral, and Mike Pompeo, former Director of the CIA. Not a smoking gun, but supportive of the lab-leak hypothesis.

9. Segreto and co-authors (working at laboratories in Austria, Canada, Japan, Spain, Australia and the USA), in a review published online in late March 2021 (in Environmental Chemistry Letters), contend that many characteristics of SARS-CoV-2 cannot be easily explained by the hypothesis that it is of natural zoonotic origin. In particular, they highlight:

   a. a low rate of evolution in the early phase of viral transmission in Humans;

   b. a lack of evidence for recombination events within the genome;

   c. a high binding affinity to human angiotensis-converting enzyme 2 (ACE2). Indeed, perhaps most importantly, the receptor binding domain of SARS-CoV-2 appears to be *highly optimised* for binding to humans;

d. the virus exhibits a novel furin cleavage site (see below);

e. the virus exhibits a flat region of the spike protein which conflicts with host evasion survival patterns exhibited in other coronaviruses;

f. despite extensive attempts to find it, no evidence for zoonotic transfer from bats or other species has been forthcoming. They review the evidence for pangolins (*Manis javanica*) being the intermediate host leading to human infection, and robustly demonstrate that this pathway remains highly unlikely. They also conclude that mink and ferrets are extremely unlikely hosts in China;

g. the suggestion that human infection occurred through contact with frozen food imported from overseas into China begs the obvious question: why were there no outbreaks of COVID-19 in these other countries exporting to China?

h. Because bats have been proposed to be natural reservoirs of SARS-r CoVs, the binding of SARS-CoV-2 to bat ACE2 should be high. However, all of the bat species examined exhibited poor rates of infection by SARS-CoV-2, and they are therefore unlikely to be the direct source for human infection. A modelling study of 38 bat species demonstrated that the binding ability of SARS-CoV-2 was either low or very low. The combination of high human adaptation and poor bat susceptibility from the first strains of SARS-CoV-2 analysed differs greatly from the evolution of MERS-CoV and SARS-CoV.

i. SARS-CoV-2 is a Sarbecovirus (a sub-genus of genus Betacoronavirus). It is the only example to contain a furin cleavage site (FCS). Furin is a protease (protein splitting) enzyme encoded by the FURIN gene. Some proteins are inactive when they are first synthesised and must have sections removed to become active. Furin cuts out these sections and activates the proteins. This cleavage site dramatically increases the human pathogenesis (the process by which an infection leads to disease) of the virus and the presence of arginine (an amino acid) at its precise location increases the efficiency of the FCS 10-fold. It is also very rare to see this amino acid at this location, being recorded in less than 4 % of all known FCSs.

j. The presence and coding sequence of a FCS is important for pathogenesis and host range. Because of these traits, the addition of a FCS into viruses has been an active topic for gain-of-function research. Sadly, the US appears to have been partially funding research in the Wuhan Virology Lab that could be viewed as research leading to gain-of-function. This was occurring with the knowledge of Dr Anthony Fauci, the Director of the National Institute of Allergy and Infectious Diseases and the Chief Medical Advisor to the President in the USA. Unfortunately, Dr Fauci did not impart this information to anyone for much of the early stages of the pandemic. Debate continues about the definition of "gain-of-function" research but US funds certainly did contribute to a 2017 paper published by the WIV reporting that researchers had isolated a coronavirus from a species of bat that could be transmitted to humans, and that reverse genetics were applied to produce new recombinants of wild bat coronavirus that do not exist in the wild and are able to replicate in human cells via the ACE2.

10. Romeu and Olle, in their pre-print paper of February 2021, conclude that the structure of the furin cleavage site of the spike protein (SARS-CoV-2 entry requires sequential cleavage of the spike glycoprotein to mediate membrane fusion) is distinct from the cleavage site of closely related SARS-CoV-2 coronaviruses. This furin site is responsible for its high infectivity, high transmissibility, and its pathogenesis. Indeed, the spike protein of COVID-19 has a distinctive structure. Although furin sites are found in many viral proteins of all types of viruses, they are not found in the genus Betacoronavirus sub-genus Sarbecovirus, which are respiratory viruses, and include SARS-COV-2. How did SARS-CoV-2, alone of all Sarbecoviruses, come to acquire this unique feature? Romeu and Olle demonstrate that a random insertion mutation is highly unlikely to be the source of this furin site. They also consider viral recombination (where another virus acts as a donor of the RNA codons defining the structure of the furin site) as very unlikely because no virus with the same codon sequence has been identified. They concluded that while manipulation by humans remains unlikely, it is not impossible.

11. Peter Daszak is an eminent scientist researching zoonosis. He is also the president of EcoHealth Alliance, which is a nonprofit non-governmental organisation that supports a number of projects related to global health

and pandemic prevention. Peter Daszak was one of the first leading scientists to dismiss suggestions that the initial outbreak of SARS-CoV-2 originated from a lab-leak. In particular, he dismissed the suggestion that it was leaked from the WIV, labelling this idea a conspiracy theory. Unfortunately it was later established that he had transferred significant research funds to the WIV, via the EcoHealth Alliance, and that *perhaps* there was a conflict of interest in his public rebuttals of the lab-leak theory. However, things got worse when he was appointed to the World Health Organization team sent to investigate the origins of the COVID-19 pandemic in China, and in particular, the WIV.

12. New evidence seen by the US Department of Energy has led it to conclude (with low confidence) that the most likely source of the virus was a leak from a research lab. Similarly, the US FBI has also concluded (with moderate confidence) that a leak from a lab was the most likely source of the SARS-CoV-2 virus. Meanwhile, former director of the US Centres for Disease Control, Dr Robert Redfield, has told the US congress that the US government probably contributed funding to gain-of-function research at the Chinese viral research lab in Wuhan.

Thus, there are two theories for the origin of the COVID-19 virus. Will we ever *definitively* know the answer? Probably not. But I think the tide has turned recently and the evidence for it being an escapee from the Wuhan lab is pretty overwhelming.

In conclusion, we have seen the following in this chapter.

A virus is a submicroscopic infection unit that can only replicate inside living cells of a host organism. Viruses are about 20 – 300 nm in diameter; a really big one might be 500 nm. While all viruses have two main components, some can have three. The two components common to all viruses are a protein shell (called the capsid) and the genetic material of the virus, that is, either DNA or RNA (deoxyribonucleic acid and ribonucleic acid). Some viruses have a lipid membrane (and associated proteins) around the protein shell and this lipid membrane (and associated proteins) is derived from (pinched off from) the host cell when the virus erupts from an infected host cell.

Viruses are not alive because they lack the array of sub-cellular structures found in living cells. They also do not contain ATP. They also don't exhibit homeostasis because they lack any metabolism within their protein coat. Therefore viruses are not alive.

The origin of COVID-19 was hotly disputed for 2 years. I think the tide has turned and the overwhelming evidence is that it is an escapee from a research lab in Wuhan, China.

RECENT DEVELOPMENTS

Since writing this chapter, reports have recently surfaced naming a Wuhan Institute of Virology scientist (Dr Ben Hu) as the first known patient (patient zero) suffering from COVID-19.

THE END

SIZE REALLY DOES MATTER. WHAT IS THE LARGEST ORGANISM IN THE WORLD? WHY ARE THE TESTICLES OF LONG-TAILED MACAQUES SOOOO VERY LARGE?

A dinosaur. A blue whale. A giant redwood tree. These are the most common contenders for the prize of the largest living or extinct organism in the world. But are any of these answers correct?

Humans are fascinated by big things. Having the tallest building in the world (currently the Burj Khalifa in Dubai at 829.8 metre-tall (2,722 ft)) is a much-sought-after title of many cities. The largest ship ever built (the Seawise Giant; 261,000 tonnes, 458 m long) and the largest man-made hole (Bingham Canyon Mine, an open-cast copper mine located 28 miles southwest of Salt Lake City about 4 km in diameter and about 1.2 km deep and visible from space) are notable achievements in size. Such man-made records, however, change with time and technology. But what of the realm of the living world? What is the largest single organism, including extinct and extant (still living) organisms? Do differences in size among different species affect any aspects of the physiology and ecology of these species? Is there such a thing as eternal life, and why are the testicles of long-tailed Macaques sooo very large? In this chapter I hope to answer these questions.

WHAT IS THE LARGEST SINGLE ORGANISM IN THE WORLD?

The first problem, as we shall see shortly, is how to define "largest". Do we mean the tallest, or heaviest, or longest/widest, or largest volume? The second problem is: what do we mean by a "single" organism? Well, I shall define the word "largest" in this context to mean any of those traits we associate with "large", including tallest, heaviest, longest/widest and largest volume. So that sorts that out. But how to define a single organism?

The Great Barrier Reef (GBR) is located off the north-eastern coast of Australia. It stretches for about 2300 km from just south of Papua New Guinea to Bundaberg, in Queensland, Australia. The GBR is the largest *living* single "structure" in the world. It covers more than 340,000 km^2. But it can't be classified as a single organism for two simple reasons. First, the reef is made up of millions-upon-millions of small animals (coral polyps) living in huge colonies that constitute a reef. The GBR probably contains 600+ species of reef forming polyps. Corals use calcium carbonate to form a hard skeleton that is white. The polyps and calcium carbonate skeleton themselves are generally transparent, with no colour. The second reason that the GBR (or indeed any reef) cannot be considered a single organism is because coral reefs rely on a symbiotic relationship between the polyps and tiny, coloured, algae (zooxanthellae). There are probably a dozen-or-more species of zooxanthellae living in a coral reef. It is the presence of algae that give corals their colour, and differently coloured algal species in different corals generate a multi-coloured reef. The algae fix carbon through the process of photosynthesis and some of this fixed carbon is given to the polyp to support its growth and respiration. Thus, coral reefs are made up of many different animal species (the polyps) and many different algal species (zooxanthellae), so no issues with stating that a coral reef isn't a single organism.

Talking about colonies (a coral reef, for example) leads to an interesting observation about ant colonies. If, like my wife, you suffer from myrmecophobia, you may wish to skip the next paragraph......

We know that ants form colonies and we know that colonies contain female worker ants and male drones and a queen. It turns out that ants can produce super-colonies as well as colonies. One large ant super-colony was discovered in Japan and "covers", or underlies, almost 3 km^2 of land, consisting of 45,000 nests, all interconnected by underground tunnels and containing more than 300 million ants, including 1 million queen ants. However, this super-colony pales into insignificance compared to a super-colony of Argentinian ants which currently exists along a 6000 km stretch of Atlantic and Mediterranean coast in Europe. This latter super-colony contains billions of ants and millions of nests, all interconnected underground. This is not a place for those suffering myrmecophobia. Obviously ant colonies (and super-colonies) are not a single organism, but those super-colonies sure are BIG.

But, back to the main question: what might be the largest *single organism* living now or in the past?

The three most common answers to this question are: a dinosaur, the Blue Whale (*Balaenoptera musculus*), and a giant redwood tree (*Sequoiadendron giganteum*). I'll take a look at the claims of each, and then provide an alternative answer.

DINOSAURS

Dinosaurs, as a group, show extreme variation in size. Perhaps the smallest was *Albinykus baatar* (Fig. 1), discovered in 2011 at Khugenetslavkant in the eastern part of the Gobi Desert. Weighing-in at less than 1 kg, this tiny dinosaur's diet may have consisted of ants and termites. Alternatively, *Linhenykus monodactylus*, found in fossil deposits in Inner Mongolia, might be the smallest, being so small as to be able to stand in the palm of your hand.

Figure 1: Life restoration of Albinykus based on the holotype and the skeletal diagram of a composite Alvaresaur by Scott Hartman. This image is licensed under the Creative Commons Attribution-Share Alike 3.0 Unported license. From: https://commons.wikimedia.org/wiki/File:Albinykus_LM.png

At the other extreme, the largest terrestrial animal ever to have **walked** on Earth is a member of the sauropod dinosaurs, a group of dinosaurs characterised as having very long necks, small heads, long tails and four thick, pillar-like legs. These herbivores are named Titanosaurs, because of their vast size. The largest may have been *Argentinosaurus huinculensis* (so-named because it was discovered in

Argentina), estimated to have weighed up to almost 100 tonnes and up to 30+ m long. Because no-one has found an entire skeleton of an individual *Argentinosaurus,* estimates of weight and length are exactly that – estimates. The more recently (2014) discovered *Patagotitan mayorum* might be longer, although possibly lighter, than *Argentinosaurus huinculensis*. These behemoths were slow-moving (hard to accelerate 75+ tonnes to a large speed on thick stumpy legs) herbivores, with eggs the size of footballs (the largest eggs in the animal kingdom). Their size conferred several advantages, though, such as some degree of protection from smaller carnivorous dinosaurs, increased tolerance to cold (because their large mass stored a lot of heat and so cooled down slowly) and a more efficient gut system able to digest more fully the plant material they ate. Unfortunately, none of this protected them from the Great Extinction of about 65 Mya (more properly known as the Cretaceous-Tertiary extinction event or the K/T extinction event). The K comes from the German word for chalk (Kreide). The Great Dinosaur extinction event is discussed in detail in Chapter 2.

But dinosaurs are not the largest organism to have *lived* on Earth, but they were the largest to *walk* on Earth. The accolade for being the largest organism to have lived on Earth might, according to many textbooks and websites, go, of course, to the Blue Whale......

THE BLUE WHALE

The Blue Whale is BIG. Indeed, it is the longest and heaviest *animal* to have ever lived, larger than the largest known dinosaur. The heaviest recorded weight (to-date) for a Blue Whale was 173 tonnes, with a length of 30 metres. For comparison, an African elephant might typically weigh about 6 tonnes. So, is the Blue Whale the largest single organism to have ever lived.....? We shall see.

GIANT SEQUOIAS AND GIANT REDWOODS ARE VERY TALL

If we take height as the principal measure of size, then the prize for the largest (i.e., tallest) organism in the world goes to the giant coastal redwood (*Sequoia sempervirens*) of California. These trees can reach heights of about 110 metres. The tallest known individual is almost 116 m tall and is called Hyperion. Hyperion means The

High One and was one of the 12 Titan children of Gaia (Earth) and Uranus (sky) in Greek mythology. The Californian coastal redwoods are endemic to a narrow band along the coast of California and Oregon. They require frequent fog cover to allow rehydration of their canopy, frequent winter rain and moderate year-round temperatures. The world's first-, second-, third-, and sixth-tallest known trees are redwoods.

If we use volume, and hence, in this instance, mass (remember that mass equals volume multiplied by density) to measure "bigness" then the prize for the largest organism on Earth is a giant sequoia (*Sequoiadendron giganteum*) of California. Although shorter than coastal redwoods (*Sequoia sempervirens*; average of about 92 m for the giant sequoias and about 112 m for the coastal redwood), the basal diameter for a giant sequoia (and the diameter of the major branches too) is almost double that of the coastal redwood. The basal diameter of a giant sequoia is up to 12 m, but for the redwood it is a *mere* 7 m. Because of these differences in diameter, the total mass of the giant sequoia can be more than 1200 tonnes compared to only 700 – 800 tonnes for the coastal redwood, because the wood density in both species are the same, so larger volume equates to larger mass. That's the same as 12 *Argentinosaurus huinculensis* all huddled together as the giant asteroid came crashing down to earth (see Chapter 2). This massive bulk takes a long time to accumulate; these trees are also some of the longest lived of all organisms. The oldest individual of this species is about 3500 years old, so it was already 1500 years old when Jesus was walking around. But this age pales into insignificance compared to some individuals of the Bristlecone pine (*Pinus longaeva*; Fig. 2) growing in the USA. One of these trees is thought to be more than 5000 years old, which means it germinated in the Bronze Age, when Stonehenge was being constructed in the UK.

CLONES, OLD AGE, AND IMMORTALITY

Since we are on the topic of old age......

Some trees are clonal – which means they reproduce asexually from pre-existing tissue, commonly, but not exclusively, roots. The death of a tree stem of Huon Pine, for example, leads to the resprouting of a new stem/trunk that grows into a tree from the original root system. Similarly, many Eucalypt species resprout from lignotubers, for example after a fire has been through a woodland. Lignotubers are below-ground storage organs beneath the main trunk. It has been estimated that some Huon pine root stocks are more than 10,000 years old. The newest tree trunk emerging from a 10,000-year old root stock isn't that age, but the root stock is. But even this age pales into insignificance when compared to the estimated 43,000-year-old (using C dating technology: see Science vol 277, issue 5325 July 1997) King's Holly (*Lomatia tasmanica*), in Tasmania, Australia. Although this is a flowering plant, it produces no seed. This species possesses three, rather than the normal two, sets of chromosomes, which makes it impossible to divide into the two equal halves required for sexual reproduction. It reproduces asexually (vegetatively) when a living branch falls to the ground and successfully produces new roots fast enough to support a new clonal plant.

Figure 2:   Examples of really old trees. These Bristlecone pines are probably more than 4000 y old. This file is licensed under the Creative Commons Attribution 2.0 Generic license. Author: Rick Goldwaser. Image from https://commons.wikimedia.org/wiki/File:Gnarly_Bristlecone_Pine.jpg

But even this amazing record is nothing compared to the "immortal jellyfish" (*Turritopsis dohrnii*; originally mistakenly identified as *Turritopsis nutricula*). Jellyfish have two stages in their life cycle: a polyp-forming colony which doesn't float around in the ocean but is attached to a substrate, and the jellyfish-like form we all recognise, called a medusa. Once the medusa has reproduced sexually, by releasing eggs and sperm, it dies. The adult has a lifespan and is not immortal because it dies. But the "immortal jellyfish" does something different to most jellyfish. When environmentally stressed, immature medusa, with no gonads, undergo a transformation from an individual with several distinct cell types (epidermal cells, stinging cells, gonadal cells, etc.) into an amorphous jelly with no cellular differentiation, which then forms a polyp, which grow, and release new medusae. This cycle bypasses the sexual reproductive stage, and thereby bypasses death. Thus, an individual is, possibly, immortal, assuming it doesn't succumb to disease or predation. I say possibly immortal because, of course, no-one has followed a single individual for an infinite length of time to demonstrate that the same individual has transformed itself an infinite number of times, thereby bypassing death.

BACK TO THE BIG QUESTION

Both sequoias and redwoods are conifers – like pine, spruce, and fir trees. Thus, they are non-flowering trees and are a more ancient lineage of plants than flowering plants. The second tallest known tree (named Centurion) is the tallest *flowering* plant in the world and is a Eucalypt, *Eucalyptus regnans* or Mountain Ash. Trees of this species rarely surpass 100 m in height. However, its diameter can surpass that of the giant sequoia (up to 15 m) and therefore the possibility exists that it is slightly heavier than the sequoias.

So, trees do seem to be the tallest **and** heaviest single organism known to biology. But are they the largest, if we define largest to mean having the largest distance between one side and the other side of the body (or the top and bottom of a body), i.e., the largest physical dimensions?

Well, not quite …

THE LARGEST ORGANISM IN THE WORLD IS A ……. MUSHROOM – YES, REALLY.

The fruiting body of a fungus is what we generally buy, chop, cook and eat, and refer to as mushrooms. Is there a difference between a mushroom and toadstool? Not in any meaningful scientific way, although colloquially, many people refer to mushrooms as edible and toadstools as inedible (poisonous). But both are the reproductive organs of a fungus. The picture below (Fig. 3) is the reproductive organ (spore-releasing mushroom) of *Phallus impudicus* – yes, really.

Figure 3: The fruiting body of *Phallus impudicus*. From https://en.wikipedia.org/wiki/Phallus_impudicus Attribution: Jörg Hempel. This image is licensed under the Creative Commons Attribution-Share Alike 3.0 Germany license.

Figure 4: A clump of *Armillaria ostoyae* mushrooms. Image from: https://en.wikipedia.org/wiki/Armillaria_ostoyae This image is licensed under the Creative Commons Attribution-Share Alike 3.0 Unported license. Attribution: Alan Rockefeller.

Fungi of the genus *Armillaria* are parasitic and infect trees and woody shrubs (Fig. 4). Some species are considered very tasty by chefs. The mycelial mat (a collection of interwoven hyphae, which are very thin, often white, hair-like filaments) and string-like structures (rhizomorphs) of most fungi grow underground and commonly extend a couple of metres (or tens of metres) through leaf litter and soil of woodlands. It is almost impossible to *visually* differentiate between the mycelia of one individual fungal body from the mycelia of another individual. Individuals are differentiated by having distinctly different DNA. But the typical "couple of metres" of mycelia of most fungi is no-where near the size of the largest single individual mushroom...... which is described in the next paragraph.....

In the late 1990's, a large area of forest in the Malheur National Forest in East Oregon died. When the Forestry service went to investigate, they found an infestation of *Armillaria solidipes* on many of the dead and dying trees. When genetic analyses were undertaken, it was discovered that two trees that were 3.8 km apart were infected by a *single clonal colony*. Indeed, a total of 61 trees were infected by this single organism. This single organism covered almost 10 square kilometres of forest floor. We can state that this is a single organism because all of the fungal cells (hyphae) were genetically identical and there was direct communication between the different hyphae (communication in terms of biochemical material moving between connected hyphae). There is also the possibility that this organism is almost 9000 years old. I think it definitively wins the prize for being the largest (3.8 km wide, covering 10 km^2 of forest floor) single organism in the world – even though most of it is unseen. It is possible one of the oldest too.

RELATIONSHIPS AMONG CHARACTERISTICS/TRAITS VARY WITH SIZE

Allometry is the study of how characteristics or traits of living organisms (characteristics such as weight, length, metabolic rate, growth rate, heart rate, leaf size) of living organisms (plants, animals, fungi, bacteria) vary with size of the organism. Allometric studies can be undertaken within a single species. For example, a comparison of the ratio of brain mass to whole body mass in humans, from birth to adulthood, demonstrates that the brain is a much larger fraction of entire body mass at birth than it is at maturity. This means that the body grows proportionally more than the brain during childhood/teenage years. This is in marked contrast to the human heart, which remains a constant fraction of total mass from birth to adulthood. Allometric studies can also be conducted across species within a class of animal (for example, comparing different species of mammals), or across major classes (for example, a comparison of traits of birds and mammals).

Some very interesting (some may say weird) relationships arise in allometric studies. I shall look at a few now, before relating some theories about "universal rules" governing allometric relationships.

SEX, AND SOME OTHER INTERESTING ALLOMETRIC STUDIES

In this section, I discuss two examples of allometric studies. The first discusses sperm competition in primates, the second discusses lifespan as a function of body size.

TESTICLES, MATING SYSTEMS, AND SPERM COMPETITION IN PRIMATES, INCLUDING H. SAPIENS

An adult male chimpanzee weighs about 50 kg and has testicles that weigh about 150 g. That is a ratio of 0.003 (150/50000 = 0.003). An adult male human might weigh about 100 kg and has testicles that weigh about 50 g, a ratio of 0.0005. An adult human male is therefore twice the size of an adult male chimpanzee but has testicles that are one third the mass of the chimpanzee's. Oh dear.

An adult male gorilla might weigh 170 kg but has testes weighing about 30 g, so his ratio is a miserly 0.000176. He is much larger than a human but has much smaller testicles. Dear oh dear...

At the other end of the scale, an adult male long-tailed Macaque (*Macaca fascicularis*) weighs around only 4 kg (about 4 % of an adult human male or 2.3 % of a gorilla), but have testes that weigh 30 g, giving a ratio of 0.0075 for the Macaque – much larger proportional to his total mass than the human, the gorilla or the chimp. They really do stand out like the proverbials......

Why might there be such a large range in the relative size of testes in adult male primates?

Well, it all comes down to sperm competition within a vagina. Or, put another way, it all comes down to the breeding systems employed within each species.

Gorillas and orangutans have breeding systems whereby one dominant male monopolises access to receptive females. Each female mates with only the dominant male (thus these species are polygynous – males have more than one mating partner, females, generally, do not). All the sperm that are racing to fertilise the egg come from one male and there is no competition from sperm of other males. In contrast, female chimpanzees, when sexually receptive, will mate with multiple males within a single cycle (these species are polyandrous). Each male now faces a problem – how to maximise the chances of his sperm winning the race to fertilise the egg. One way to increase his chances to pass on his genes to the next generation is to deposit a very large number of sperm. To produce and store a large number of sperm, the male needs large testicles. Because monogamy (one male mates with one female partner throughout any given breeding season or across multiple/all breeding seasons) is a breeding system where sperm competition among males is absent, it follows that monogamous species should have relatively smaller testes than species where sperm competition is present. This seems to be the case (Fig. 5). Species exhibiting sperm competition (multiple males copulate with a single female within a single cycle of ovulation) have, on average, larger testes, per unit body mass, than species without sperm competition (monogamous species and species where a single dominant male copulates with all the receptive females, i.e., polygynous). This means that for most species (i.e., where body mass exceeds one kilogram), the size of the testes at any given body mass (1 to 100+ kg) are larger for species where sperm competition occurs.

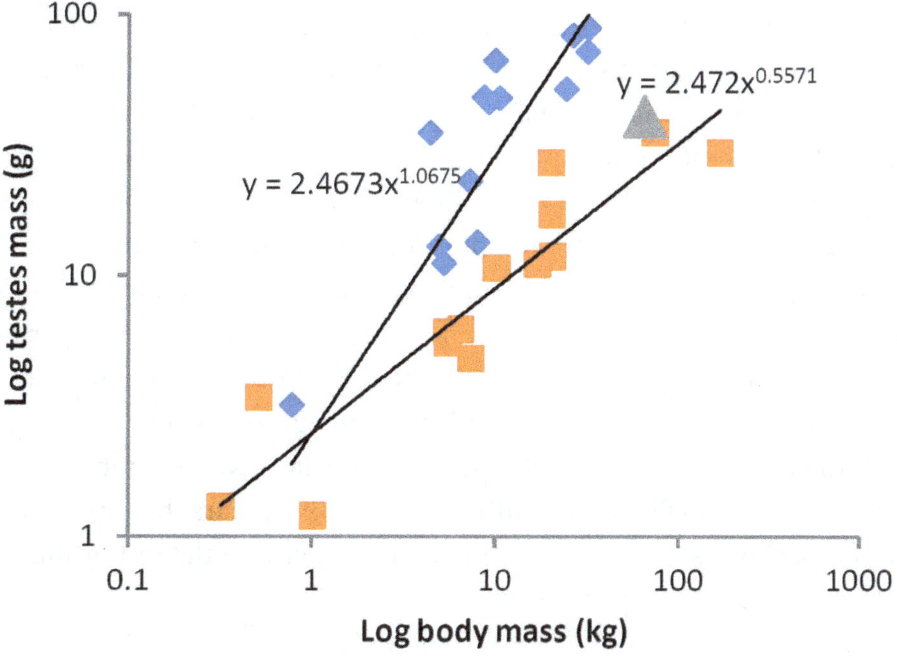

Figure 5: A log-log plot of body mass *versus* testes mass for multiple primate species. The blue diamonds represent species where sperm competition is present; orange squares represent species where sperm competition is absent. The green triangle is *H. sapiens*. Humans, it appears, have evolved in the absence of significant sperm competition. Redrawn from Harcourt et al., 1981, Nature 293.

Two additional strategies to winning the sperm competition race have been identified. The first is to have sperm that swim faster than your competitor. Sperm have three regions: a head, containing the nucleus where the DNA required to fertilise the egg, is stored. A mid-piece, or connecting-piece, that is packed with mitochondria, and the long thrashing tail (the flagellum) that propels the sperm. Swimming by sperm takes a lot of energy and the chemical energy driving the thrashing of the tail is produced by the mitochondria, the sub-cellular structures that produce chemical energy stored in the ubiquitous "energy currency of cells" – namely ATP (adenosine triphosphate; Chapter 7). It has been hypothesised that in species where sperm competition is present, the mid-piece should be larger, thereby providing more energy to the tail, to assist in the inter-male sperm competition. In 2002, Anderson and Dixson demonstrated that across 31 species of primates, the average volume of sperm mid-pieces of species with sperm competition was almost double (9.8 μm^3 compared to 5.4 μm^3) the volume in species where sperm competition was not occurring, supporting the hypothesis that natural selection of males of species with sperm competition resulted in sperm with a larger store of mitochondria. Interestingly there appears minimal variation in the total size of the sperm cell across species. But talking about sperm size, although proportionally-speaking, the longest known sperm cell (relative to body length) is produced by a small fruit fly – *Drosophila bifurca*. While only about 3 mm long, an adult male fruit fly produces a sperm cell that is almost 60 mm long, or 20 times longer than its body. An equivalent length for a human sperm would be almost 40 m long!

The second strategy to improve the probability of winning the sperm-war is to insert a plug into the vagina of females after ejaculation, thereby reducing the chance of sperm of a second male fertilising the egg. Although the plug can be ejected by the female after a period of time, that delay gives a competitive edge to the sperm of the first male mating with the female. It appears that this strategy features more in primates with significant sperm competition than in primate species where sperm competition is absent.

HAVE HUMANS EVOLVED TO A PROMISCUOUS OR A MONOGAMOUS STRATEGY?

From Figure 5 we see that the human testes/body weight ratio is very close to that of monogamous primates, which suggests a monogamous evolutionary history (but see below for further discussion of this). In addition, the relatively simple structure of the human penis, lacking as it does cone-like protuberances, flanges, and other structures, is more like the penis of monogamous primates than those of chimpanzees, for example. But humans do appear to invest in the so-called "sneaky fuckers" syndrome (or kleptogyny, as it is scientifically known). In gorillas and deer, both of which maintain a harem of females, the dominant male spends a significant amount of time in the mating season fighting off other males and protecting his harem. However, when the dominant male is warding off another male making a challenge for access to the females, some sneaky males use this time to lure a female away and mate, out of sight. It is a risky strategy, but the rewards for the non-dominant male can be very large – making a female pregnant increases the chance of his genes passing on to the next generation. While some estimates put as many as 25 % of human children having a biological father other than the partner/husband of the mother, a more realistic estimate is probably 2 – 3 %. So even in humans, the sneaky fucker strategy appears to be at play, despite monogamy being the most common pattern among the proletariat (as distinct from the very wealthy/powerful/high status males of society).

The idea that monogamy (a single mate for both male and females of a pair; this might be for one breeding season or many/all breeding seasons) is the "natural" or the most common mating system for humans is contentious. In a review of human cultures around the world, Bernard Chapais (2013) stated that, compared to that of other primates, the human mating system is extremely flexible, and embraces short-term and long-term mating systems that can be either overt, with either monogamous or polygynous structures (but rarely polyandrous), or covert, i.e., with social disapproval. He contends that in overt mating systems (where everyone knows what is going on) within human societies, polygyny and monogamy are dominant in 80+ % of human societies, and strict monogamy is enforced in only 17 % of societies. However, monogamy is the absolute dominant form of mating system for most men globally, because few men have more than one wife at the same time, and extremely few women have more than one husband at the same time, globally. However, the occurrence of covert mating systems (where few outside the pair know of the pairing) is far more variable across societies.

The reason for the dominance of monogamous behaviour (including serial monogamy) in humans is highly debated. The list of reasons include (but is not limited to): preventing infanticide by non-father males; maintaining exclusive access to a reproductive mate (note that humans do not have a breeding season, unlike other primates); the need for two parents to rear an offspring that is incapable of independent life for very many years; division of labour to maximise the probability of survival of offspring; increased geographical spread of females made it increasingly difficult to guard/protect multiple females simultaneously).

BIGGER ANIMALS LIVE LONGER THAN SMALLER ANIMALS

Bacteria are very small, typically 1 – 10 microns (0.001 – 0.01 mm). They don't live long before dividing to produce two progeny. Bacterial lifespans are measured in hours. Fruit flies are a couple of millimetres long and live about 40 days. Elephants weigh 4 – 6 tonnes and live about 50 – 70 years. Figure 6 shows the relationship between the natural log of body mass and natural log of lifespan for a number of mammal species. On average, bigger species do live longer than small species. Teasing out why has been quite an effort.

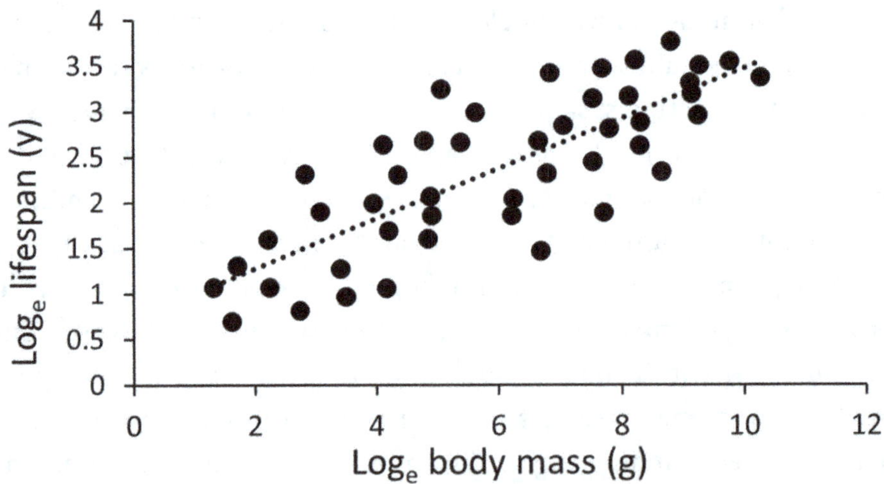

Figure 6: The natural log (Ln) of body mass and Ln lifespan for a number of mammal species. Redrawn from Speakman 2005. Natural logs (as opposed to log10) is used to make the relationship linear, which makes it easier to compare across data sets.

Aristotle thought that the more sex that animals undertook, the earlier they died. I shall leave that thought with you.

> **Natural Logs**
> The **natural logarithm** of a number is its logarithm to the base of the mathematical constant e, which is a number proximately equal to 2.718281828459.
>
> The natural logarithm of x is generally written as ln x, or $\log_e x$. Parentheses are sometimes added for clarity, giving $\ln(x)$ and $\log_e(x)$.
>
> The natural logarithm of x is the power to which e would have to be raised to equal x. For example, $\ln(7.5)$ is 2.0149..., because $e^{2.0149...} = 7.5$. The natural logarithm of e itself, $\ln(e)$, is 1, because $e^1 = e$.

An early theory to explain mortality of animals was the effect of "wear-and-tear". Perhaps large species somehow exert less wear-and-tear on their bodies and so live longer? Early in the 20th century, Max Rubner measured the rate of metabolism (presumed to be a measure of wear-and-tear) of six species, including guinea pigs, humans, and horses and saw that the rate of metabolism increased with size, and that lifespan increased with size. However, when he calculated the mass specific rate of metabolism (energy consumed per kg of mass) and multiplied this by lifespan, he found it to be more-or-less a constant (if data for humans were ignored; human lifespan by this time was benefiting from the benefits of a range of technologies and medical interventions, which skew the data). This means that each kilogram of body mass expends the same amount of energy before the animal dies. Therefore, if large species can reduce the rate at which they consume energy at a cellular scale, then they should live longer than smaller species (because tissues consume the same amount of energy before they die, independent of species size). This is known as the *Rate of Living Hypothesis*, summarised by "Live Fast, Die Young". A biochemical mechanism for this hypothesis arose when free-radicals were identified as causing damage to membranes, proteins, and DNA, in the 1950's. Free radicals (reactive molecules of oxygen whereby there is an extra electron attached, temporarily, to an oxygen molecule) are produced in abundance during sub-cellular respiration. Although cells have protective processes to "mop-up" these free radicals, they can still cause damage within cells, and the more oxygen consumed, the faster the rate of production of free radicals. Of course, as the rate of metabolism increases, the rate of oxygen consumption and therefore the rate of free radical production, increases.

When we multiply daily energy expenditure by lifespan and express daily energy expenditure per unit mass, thereby removing the confounding effect of comparing small with large animals, we find a strong negative correlation for mammals and a poorer, but still negative and significant correlation for birds (data not shown; Fig. 7). A negative correlation simply means that as the value on the x-axis increases (body mass in this example), the value of the y-axis (metabolic rate per gram tissue per lifetime in this example), decreases.

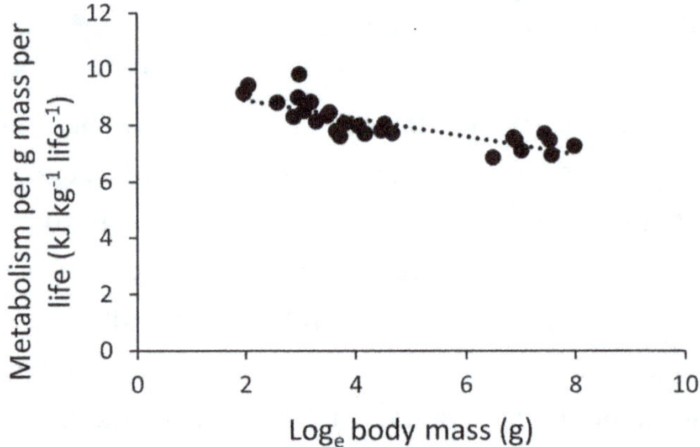

Figure 7: Lifetime energy consumption per gram of mass declines with increasing body mass. These data are for a range of mammal species. Because the linear regressions slope downwards, this is a negative correlation.

It appears, therefore, that larger species have a lower metabolic rate per unit body mass per lifetime and this would appear to support the free radical idea and hence the rate of life theory (it is called the rate of life theory because as the rate of metabolism increases (i.e., the rate of living increases) the rate of production of free radicals increases). By having a lower rate of metabolism per unit mass, the rate of damaging free radicals would appear to be reduced, leading to a longer life. If only it were this simple.

The idea that rate of living and lifespan are inversely related, i.e., that a shorter lifespan results from a faster rate of living (the rate of life theory) and that this generates more reactive oxygen molecules that damages cells (i.e., membranes, proteins, DNA), held sway for most of the 20th century. While problems with small data sets and the confounding effects of body mass were often ignored, more recent analyses using larger data sets, and accounting for confounding effects, show:

1. For birds and mammals (at least), larger animals do indeed live longer,
2. For birds and mammals (at least), lifetime energy expenditure per gram of tissue declines with increasing body mass,
3. The lifetime accumulation of damage arising from reactive oxygen species produced in mitochondria remains a viable hypothesis to explain why larger animals live longer.

These relationships can be summarised mathematically by defining the equation linking changes in the variable of the x-axis (for example, body mass) to changes in the variable of the y-axis (for example lifespan). These mathematical equations are given the generic term "scaling relationships (see text box below) and because there appears to be some strong commonality among all the equations (see below) this has been expanded into the phrase "Universal scaling laws". In the next section, I explain what this means……

Is there a universal scaling law in biology?

There does seem to be a universal scaling law and the scaling exponent (see text box on Scaling Laws) appears to be $1/4$ or multiples of $1/4$ (including negative values). These scaling laws are given the name "quarter-power allometric scaling laws". This idea is most forcefully proposed by Geoffrey West, Brian Enquist, and their colleagues, although there is an opposing view that perhaps the scaling exponent is $1/3$ and not $1/4$. I don't want to enter that argument, but I shall present some reasons why I think the $1/4$ is more likely to be correct.

What are scaling laws?
What is a scaling law? A scaling law is the mathematical relationship between two variables, for example, a measure of body size (e.g., mass) and another trait, for example lifespan or metabolic rate.

A simple example of a scaling law is the relationship between the volume of a cube and the length of the side of the cube. A cube that is 10 cm x 10 cm x 10 cm has a volume of 1000 cm^3 (i.e., 10 x 10 x 10). A cube that is 15 cm x 15 cm x 15 cm has a volume of 3375 cm^3. The scaling relationship here is length

(L) raised to the power of 3 (i.e., L^3; i.e., in the examples just given, 10^3 or 15^3). Hence scaling laws are also called power laws. The number 3 is the exponent: it is the number that is written as the superscript; it is the power by which the variable (length, mass, etc.) is raised. The power law, scaling relationship between volume (V) and length (L) of the side of a cube is written:

$V = L^3$

The scaling exponent here is 3.

The scaling exponent relating to the area of a circle and its radius (r) is 2, because the surface area (SA) of a circle is given by πr^2 (SA = πr^2).

Just because the exponent is a constant in the equation, doesn't mean the magnitude of the increase in the y variable (for example, the volume of a cube or the surface area of a circle) for each unit increase in the x variable (for example, the length of the side of the cube or the radius of a circle) is constant. It isn't, because the relationships between the y variable and the x variable isn't linear, as demonstrated in Figure 8. For each unit increase in radius or length, the increase in surface area or volume gets larger and larger.

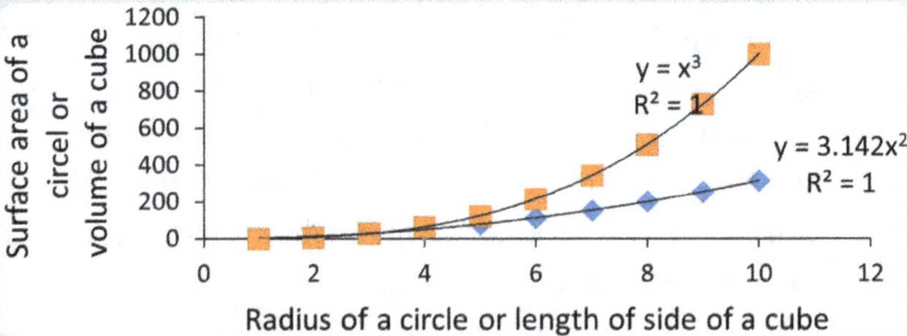

Figure 8: The relationship between radius of a circle and surface area (blue diamonds), or L and volume of a cube (orange squares), are not linear.

How to make curved lines on graphs into straight lines and the meaning of exponents

We can often (not always) make curved relationships look linear by plotting the natural log (represented as Ln) of the x and y variables. The natural log of a number is the logarithm to the base e, where e (like π in geometry and trigonometry) is a mathematical constant (approximately equal to 2.71828) of x and y and do the linear regression, the **exponent becomes the number preceding "x".**

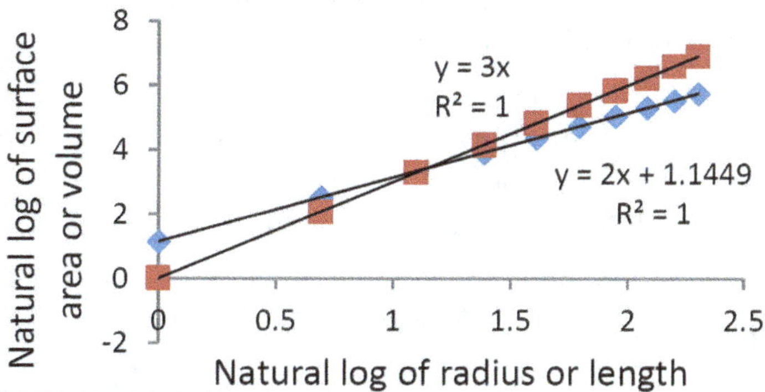

Figure 9: The relationship between radius of a circle and surface area (blue diamonds), or L and volume of a cube (red squares), can be made linear by plotting the natural log values. The exponent of these two lines are 2 (blue data) and 3 (red data) because the exponent is the number preceding the x in the linear equation.

Exponents don't have to be whole numbers; they don't even have to be positive values. It is possible to have an exponent value of ¼ (i.e., 0.25) or -¼ (i.e., -0.25), for example. When a number has an exponent that is a fraction, for example ½ (i.e., 0.5), this means to take the square root of the number (so $9^{0.5}$ means $\sqrt{9}$, which =3). Similarly, $27^{1/3}$ means the cube root of 27 (=3). When the exponent is ¼ this means take the fourth root, so $81^{0.25}$ equals 3 (3 x 3 x 3 x 3 = 81). If the fraction is negative (e.g., $27^{-1/3}$) then this means $1/27^{1/3}$ which is 1/cube root of 27 = 1/3 = 0.333. The negative means that you take the inverse (i.e., one over) of the number, and then deal with the exponent.

Turns out that if there is a universal scaling law, then the exponents seem to be ¼ or multiples of ¼ (or negative ¼ and multiples).

In 1932 Max Kleiber published data that linked daily metabolic rate to body mass. The log: log plot had a slope of 0.74, so the exponent is ¾. Which is a multiple of ¼. His data consisted of 13 data points across 10 species. The range of body mass covers almost four orders of magnitude, but Geoffrey West increased this range to 27 orders of magnitude by including more animals, including an elephant but perhaps more impressively, by including unicellular organisms and sub-cellular organelles, such as mitochondria and even individual molecules involved in sub-cellular respiration (Fig. 11). When this was done, he obtained the ¾ power exponent, powerful evidence that this universal scaling law is applicable across a gigantic scale – from a 6-tonne elephant to a protein weighing about 10^{-18} g (or one millionth of one millionth of one millionth of 1 g).

A quick note on regression lines

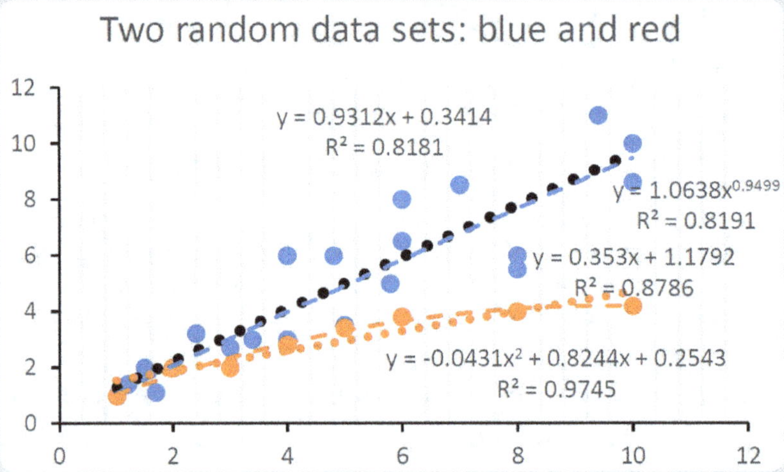

Figure 10: A regression line is merely a line calculated using a statistical process (regression analysis) and this process can determine the best fit of many different types of curves, including linear, polynomial, or a logarithmic curve, to a data set. The statistical process tells us the equation that best describes the relationship between two variables and give us a measure of how well the line fits the data. If the regression coefficient (R^2) equals 1, the regression fits the data perfectly. For the orange data points, I have fitted a polynomial curve (dashed line) and the $R^2 = 0.9745$, meaning the line is a very good description of the data. A linear regression (dotted line) fitted to these data yields an $R^2 = 0.8786$, which is good, but not as good a fit as the polynomial curve.

For the blue data, the linear regression (black dots) yields an $R^2 = 0.8181$ and this reflects the wide spread of data around the regression line. The power curve (dashed line) yields an $R^2 = 0.8766$, which is better than the linear regression, even though the two lines are almost on top of each other.

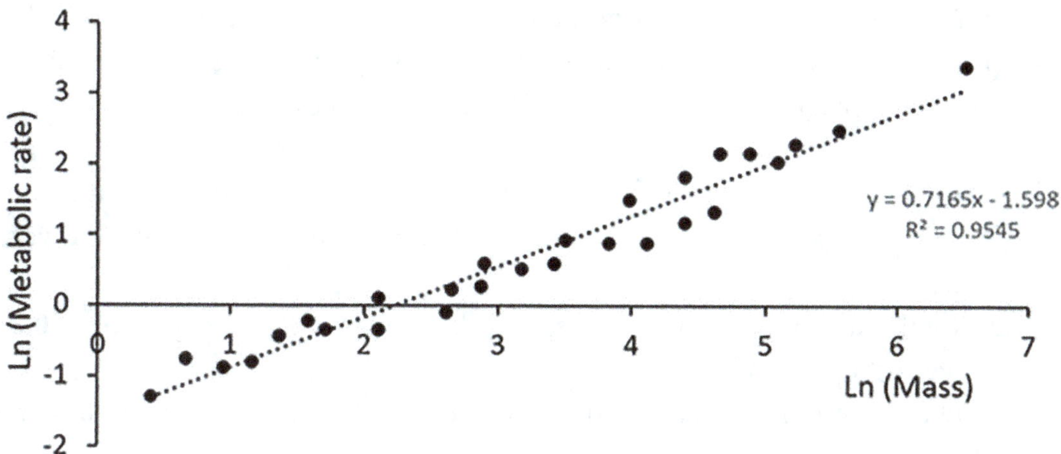

Figure 11: Relationship between natural log of mass *versus* the natural log of metabolic rate (or metabolic power) in mammals (upper panel) and unicellular organisms and molecules (lower panel). The exponent in both is very close to ¾ indicating a 3/4-power law for the metabolic rate as a function of mass. Redrawn from West and Brown 2005.

Another example of the ¼ (or multiples thereof) scaling law can be found in the relationship between pulse rate and body mass across 17 species, from a canary (mass about 20 g) to an elephant (mass about 5000000 g; Fig. 12). Here the exponent is -0.274 and this is not statistically different from -¼.

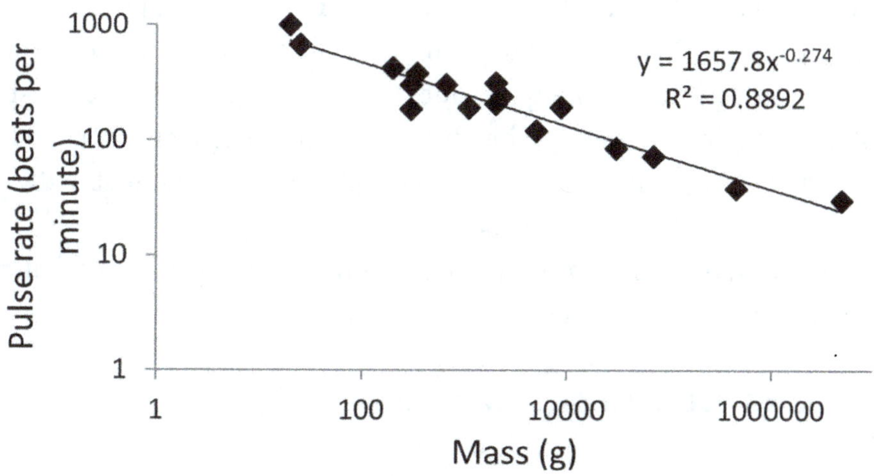

Figure 12: The relationship between body mass and pulse rate for 17 species of birds and mammals, plotted on a log:log scale (hence the linearity). The exponent here is 0.274, which is very close to ¼.

It turns out that there is a large number of relationships that scale to ¼ (or multiples) and this scaling is applicable to biochemistry, physiology, ecophysiology, and micro- and macro-ecology. Examples include the following:

1. Annual rates of biomass production (i.e., growth rates) of unicellular and multicellurar plants scales with body mass with an exponent of 0.76 (very close to ¾).
2. Total cell pigment content scales with body mass with an exponent of 0.72 (very close to ¾).
3. Photosynthetic biomass scales with non-photosyntetic biomass with an exponent of 0.762 (very close to ¾).
4. The number of trees per hectare of land scales with the aboveground biomass of the trees with an exponent of -0.75 (i.e., -¾).
5. Genome length scales with cellular mass with an exponent of 0.24 (very close to ¼).
6. Pulse rate scales with body mass with an exponent of -0.27 (very close to ¼) (Fig. 12).
7. Height of a tree scales with mass of the tree with an exponent of 0.25 (i.e., ¼).

It is truly remarkable that the exponent ¼ (or its multiples), applies across spatial-scales ranging from the size of a protein molecule to forests, across temporal scales ranging from seconds to many decades, and ranging across such disparate fields of biology as molecular biology, biochemistry, physiology, anatomy and ecology, and across the plant, animal, fungal and eubactaria, and presumably the other two kingdoms too (protists and archaebacteria).

The question now, of course, is: are there any fundamental principles that might explain the existence of such a universal scaling law?

WHY DOES SCALING FOLLOW A ¼ POWER LAW?

All individuals and populations of individuals, both intraspecific and interspecific, are unique – unique in terms of genetic make-up, life history, size, environmental conditions, geographic distribution. Despite this, the existence of universal scaling laws suggests some fundamental mechanisms act on all individuals at all spatial and temporal scales.

All organisms, from the smallest to the largest, depend on the integration of many subunits, including molecules, organelles, cells, tissues, and organs. Each of these requires inputs of substrates (for example, amino acids and sugars) and energy, and the removal of waste. Transport (through vascular systems or by simple diffusion) and communication (for example, via hormones) are key to this. Two researchers, Geoffrey West and James Brown, propose that natural selection resulted in hierarchical, fractal-like branching networks, which distribute energy and materials among subunits. Examples of such branching networks include animal circulatory, respiratory, renal, and neural systems, plant vascular (the xylem and phloen that transport water and sugars around a plant) systems, intracellular networks, and the distribution systems that supply food, water, and power to houses.

West and Brown proposed that the quarter-power allometric scaling laws reflect the constraints (or properties or principles) inherent in the generic properties of these networks. These properties/principles include: (i) networks are space-filling, and thereby minimise the distance for diffusion from the network to the cells requiring resources; (ii) the terminal units of the network (alveoli in lungs, capillaries in blood transport systems) are (near) identical within each network; and (iii) networks are optimised by minimising the energy and substrates required to distribute materials to the end-points.

By applying these principles, West and Brown were able to model the mammalian circulatory and respiratory systems and plant vascular systems and discovered the characteristic quarter-power exponents, even though the anatomy and physiology of the pumps and plumbing are very different (trees *versus* humans, for example). Seems like a winning theory, to me.

In conclusion, we started with a simple, almost child-like question: what is/ was the largest living organism in the world? Having established what we mean by largest, and having defined individual or "single organisms" we discovered the rather surprising answer that a fungus is the largest individual organism in the world (largest here means the largest single dimension of any single organism). After metaphorically picking ourselves up from the floor, we embarked on a rollercoaster ride to contemplate mating behaviours of apes and how this determines the sizes of testes across different hominid species, including humans. The size of testes in mammals is generally related to the mating system (monogamous, polygynous, polyandrous) used by each species. Gorillas and orangutans have a mating system where a single dominant male copulates with all receptive females, and sperm competition is minimal. Therefore, these species do not need to produce copious amounts of sperm. In contrast, adult female long-tailed Macaques (*Macaca fascicularis*) mate with multiple males during each mating season. Each male Macaque needs to therefore produce copious quantities of sperm, to try and outcompete the sperm of the other males, hence male Macaques have massive testes relative to their body size.

This naturally led us to think about allometric relationships – the study of how relationships among traits vary with growth/size – which finally led us to conclude that there are some universal scaling rules that are applicable across vast spatial scales, from molecular, whole tissue, organ, and population/ecosystem-scales. Examples of relationships conforming to the universal scaling law includes:

- body mass and pulse rate
- bigger animals live longer than smaller animals
- metabolic rate of mammals per gram of tissue per life declines with increasing body mass
- heart rate and lifespan
- the number of trees per hectare of land and the above ground biomass of the trees.

Perhaps the most remarkable insight is that there is a universal scaling law that appears to exist across the entire range of sizes, from molecules, through to bacteria, small mammals, big mammals small plants, huge trees, and ecosystems.

In addition to their work on establishing scaling relationships, Brown and West were also able to establish that a couple of constraints on how networks fill 3-D space were sufficient to result in the self-organising emergence of the quarter-power relationship.

Before I end, I want to present this interesting little teaser for you to ponder: Can it be true that mammals all have essentially the same number of heartbeats in a lifetime, that is, about 1 billion (10^9) (Fig. 13)?

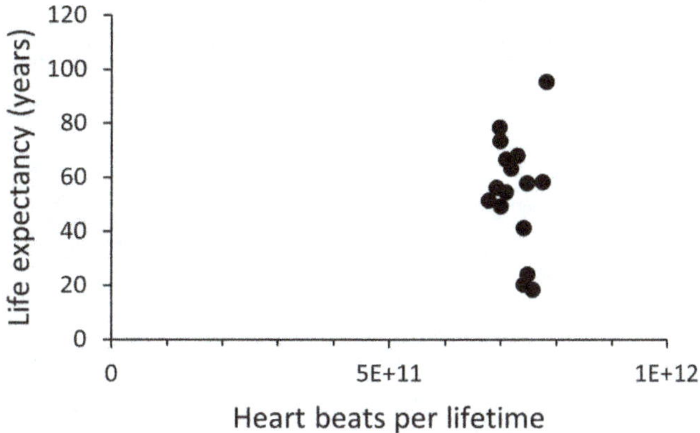

Figure 13: Heartbeats in a lifetime for different mammals. Note the $\log_{10}$ scale for both axes. Despite the large range of life expectancies the number of heartbeats per lifetime varies only slightly. The uppermost data point is for humans, the lowest data point is for mice.

THE END

CHAPTER

HOW DID LIGHTNING SIMULTANEOUSLY KILL 62 CATTLE IN A FIELD IN 2005? WHAT MAKES LIGHTNING, AND HOW IS LIGHTNING GLOBALLY IMPORTANT?

Lightning – we have all seen it and perhaps many of us were a little-bit scared when we were children, either of the lightning itself or the following thunder, or both. But even as adults, few would know how lightning can kill dozens of cattle in a single field without physically striking them. And few would realise the global importance of lightning. In this chapter I address the three questions posed in the title. In doing so, I touch upon lightning, wildfires, ozone production, ball lightning, and St. Elmo's Fire. A veritable smorgasbord of topics.

HOW LIGHTNING KILLS CATTLE WITHOUT TOUCHING THEM

Images of multiple cattle lying dead on the ground after an electrical storm are not that uncommon in rural areas globally. And when you look closely, there are no scorch marks on the hides of the cattle, in marked contrast to the appearance of trees that have been hit by lightning, which invariably show large black scorch marks running from the initial point of contact, down to the ground. And how likely would it be for a single bolt of lightning to hit multiple cattle simultaneously, or for multiple bolts to hit 10 cattle spaced across a 10000 m^2 field? Indeed, the Guinness world record for the largest number of cattle killed by a bolt of lightning is 68 Jersey cows killed in Dorrigo in northern NSW, Australia, in spring 2005. It turns out that a single bolt of lightning can kill multiple cattle, simultaneously, without actually directly hitting them…… let me explain.

Lightning bolts are typically of the order 30 million to 300 million volts. When they hit the ground, they can induce an electrical field of about 100,000,000 V m^{-1}. That means there is a difference of 100,000,000 V between where the lightning hit the ground and any point on the ground 1 m away. If a cow is facing the point of contact, and its front and back legs are 2 m apart, the difference in voltage between the front and back legs *could be* (it depends on the distance from the lightning strike to the cow and soil characteristics, including soil water and salt contents) up to 1,000,000 V. This will drive a massive electrical current through the body of the cow, killing it instantly, without the lightning bolt ever actually hitting the cow directly. This is how a single lightning bolt can kill 2, 10, 20 or 68 cows in one go, if their front and back legs are a different distance from where the lightning hit the ground. They all experience a massive electrical current flowing between their front and back legs, causing death. Sad, but unavoidable!

So, how is lightning generated?

What makes lightning?

Globally there are about 1.5 billion lightning flashes per year, or more than 40 per second, with about three quarters occurring in the tropics. More than 2000 people are killed by lightning strikes every year, although some unfortunate people (or should that be fortunate??) have been struck several times and lived to tell the tale each time.

Figure 1: Probably the most amazing photograph of lightning you are ever likely to see. Photo taken by Maxime Raynal from France. Port and lighthouse overnight storm with lightning in Port-la-Nouvelle in the Aude department in southern France. From: https://en.wikipedia.org/wiki/File:Port_and_lighthouse_overnight_storm_with_lightning_in_Port-la-Nouvelle.jpg

Lightning is associated with clouds: it is in clouds that the processes required for the generation of lightning, take place. Clouds are mostly, but not entirely, made of water droplets. As water-laden warm air rises from the ground it cools and water condenses around dust, aerosols and bacterial/virus particles, to form water droplets. We are seeing the beginnings of a cloud. As it cools further, some droplets will freeze to form small ice crystals. *Some* of these crystals are relatively hard and exist as hail particles (called graupel) and tend to be a little larger and heavier than smaller ice crystals. The small ice crystals are lighter and have a low density and consequently move upwards in the updraft within the cloud. The graupel, being heavier than water droplets and small ice crystals, tend to move downwards, under gravity. The ice crystals repeatedly bump into water droplets, other ice crystals, and the graupel. This bumping creates friction and generates static electricity in the same way that rubbing an acrylic rod with a woolen cloth can generate static electricity on the surface of the rod. The friction in clouds is sufficient to generate positively- and negatively-charged ice crystals, graupel, and water droplets, because of the movement of electrons (which have a negative charge) from one crystal or droplet to another. The donor of the electron tends to be the small ice crystals which become positively charged (having lost one negatively charged electron) and the receiver, usually the graupel, becomes negatively charged (having gained one electron). This effect (called the triboelectric effect) also occurs when aircraft fly through clouds. The friction between the ice crystals and the aircraft's "skin" generates static electricity. The faster the plane is flying, the larger the static charge produced on the aircraft's wings and fuselage. Fortunately, this charge build-up doesn't seem to cause any serious problems for the plane or its occupants. Planes are also very effective Faraday cages, whereby the fuselage protects the inside (the electronics and people) from external electrical fields.

As the cloud forms and more bumping occurs, the lower portion of the cloud tends to become negatively charged (the graupel falls downwards under gravity) and the upper portion positively charged (ice crystals float upwards). A battery, or perhaps more correctly, a capacitor (a capacitor is something that stores electrical energy in an electrical field by accumulating positive and negative charges on two opposing surfaces), is produced.

It is not well-known that the charge on the lower surface of the cloud induces accumulation of an equal and opposite charge (so, a positive charge) on the ground immediately below the cloud. The air between the base of the cloud and the ground surface is a very good electrical insulator so very large voltage differences can accumulate between the cloud base and ground surface. However, once this voltage is large enough, it can be discharged, or "jump" from cloud-to-ground, as lightning bolts. Lightning bolts can exhibit voltages of tens-of-millions to 300 million volts. Not all lightning strikes are cloud-to-ground. Many are cloud-to-cloud, and even cloud-to-air, but being human and mostly confined to land surfaces, we tend only to worry about cloud-to-ground strikes.

The process of generating a visible lightning bolt starts with an invisible process. The first step is the formation of a bidirectional channel of ionised air, called a "leader". When the voltage between the cloud and ground is large enough the electrical field is large enough to cause the breakdown of gas molecules in air. This results in the formation of a channel of ionised air and this channel is very conductive to the flow of electricity. A lightning bolt is of course a very very large flow of current, or, in other words, a very very large spark. The positively charged tip of a leader moves upwards towards the negative region of the cloud and the negatively charged tip can move down towards the ground, which has a positive charge. Each leader tip can branch repeatedly and many branch ends do not reach the ground. Leaders often have a "stop-start" process rather than moving in one continuous direction. It takes tens-to-hundreds of amps of current flow to establish these channels of ionised air and the exact mechanism by which they come about remains poorly understood. To put this value in context – about 150 milliamps (i.e., 0.15 amps) is enough to kill a human.

These leaders can be seen by the naked eye but do not constitute what is generally known as the "bolt of lightning" beloved of makers of Dracula movies. They are slower moving than the bolt of lightning. As the leader approaches the ground from the cloud, the size of the electric field close to the ground increases and this can induce formation of an upward streamer, which is the channel of ionised air moving up to meet the downward-traveling leader. Once the streamer and leader have met, a low resistance pathway (low electrical resistance, which means any lightning discharge will follow this path) is formed and a lightning "bolt" or discharge occurs. This bolt of lightning travels from the ground to the cloud and is much brighter and travels faster than the leaders. The downward leader may travel at about 200,000 mph (300,000 kph) but the return bolt of lightning travels at speeds of up to 200,000,000 mph (300,000,000 kph), or about 28 % of the speed of light. The amount of current in a single bolt of lightning is typically 30,000 amps and this is so large and the discharge so fast that is superheats the air to about 50,000 °C. Remember that the surface of the Sun is about 6000 °C. This superheated air is a plasma at this temperature and the wavelength of light emitted is "electric blue-white". A plasma is one of the four fundamental states of matter (the other three are solid, liquid and gas). The thing about plasmas is that they contain approximately equal numbers of positive and negative ions, and, in the case of lightning (and neon signs used in advertising), some neutral particles too. Because of the positive and negative ions present, a plasma is very electrically conductive. Weirdly, plasmas are the most common form of ordinary matter (i.e., excluding antimatter) in the Universe and is very abundant in stars, including our Sun. Plasma TVs use small cells, each of which contains a plasma.

Figure 2: Map showing the average yearly counts of lightning flashes per square kilometre across Earth, from 1995 to 2013. Areas with the fewest flashes each year are gray and purple; areas with the most lightning flashes (up to as many as 150 per square kilometre per year) are bright pink. NASA Earth Observatory image by Joshua Stevens using LIS/OTD data from the Global Hydrology and Climate Center Lightning Team. From: https://earthobservatory.nasa.gov/images/85600/global-lightning-activity

It is the extremely fast heating to extremely hot temperatures of the air in the return strike that causes the air surrounding the lightning bolt to expand outwards extremely rapidly. This shockwave of high air pressure travels away from the lightning bolt and is heard by us as thunder. As the air cools and contracts it causes further shockwaves to move outwards giving the rumble after the initial "crack" of thunder. If your count the number of seconds between you seeing a bolt of lightning and hearing the first clap of thunder you can estimate the distance between you and the lightning. Five seconds corresponds to about 1.6 km.

THE GLOBAL IMPORTANCE OF LIGHTNING

NITROGEN, LIFE AND LIGHTNING

Lightning is globally very important. Strange but true. The first example of its importance is the production of oxides of nitrogen (NO and NO_2) that can be used by plants. Nitrogen (N) is a major component of all proteins in living organisms, including enzymes, and therefore, for example, all of the enzymes involved in photosynthesis (see text box below). DNA contains a lot of N too. So, N in a form that is available to plants, and hence food production, is central to life on Earth. Most plants are unable to use nitrogen gas (N_2) in Earth's atmosphere because the three bonds between the two N atoms in a single molecule of N_2 are extremely stable and unreactive. Therefore plants are dependent on soil N. However, globally, many soils, including deserts, most Australian soils, and poorly-managed cropland soils, are N deficient. This is why the world is using ever-increasing (by about 2 % per year) amounts of N fertiliser, currently about 120 million tonnes per year. The production of nitrogenous fertilisers, containing ammonia, is very energetically expensive (see text box on Haber industrial chemical process).

Photosynthesis and life on earth
It is the photosynthetic uptake of carbon, as CO_2, by plants that supports all (well, 99.999 %) of life on Earth. In the absence of photosynthesis, all vegetation dies, and therefore all of the 99.999 % of life, i.e., all animals (including humans), fungi, and bacteria, dependent on that vegetation, dies. The possible exemption to such an extinction is a small (in spatial extent and total

biomass of organisms) community of bacteria, giant clams, tubeworms, and shrimp that live near deep-sea hot water springs along rifts between tectonic plates. These hydrothermal vents were only discovered in the late 1970s and the energy supply to their entire ecology is dependent on the sulphur emitted from these vents. Photosynthesis can't occur here because there is no light – it is far too deep for light to penetrate to these vents.

Some plants, including the common garden pea, runner/string beans, and *Acacia* trees, contain bacteria in nodules attached to their roots which 'fix' the N_2 in the atmosphere to produce a form of nitrogen that can be used by plants (ammonia). In addition to the fixation of N by legumes, and the industrial fixation of N as ammonia for fertilisers, lightning produces very significant quantities of oxides of nitrogen in forms that can be used by plants.

The Haber Process: industrial N fixation

The Haber process combines gaseous N_2 from the atmosphere with hydrogen (from methane) to produce ammonia. The reaction proceeds under high pressures (about 20 MPa, or 200 atmospheres of pressure) and high temperatures (about 400 °C) and requires a catalyst. Because of these requirements, it is a very energetically expensive process. However, it is the best method for the industrial-scale production of ammonia, which is required for fertilisers. It was the method used by the Germans in the production of explosives in the First World War.

As detailed previously, a lightning bolt superheats the air immediately around it to extremely high temperatures. This superheating results in a large array of chemical reactions involving N_2, N, NO, N_2O, O_2 and O_3 (ozone (see below)), and, significantly, produces large quantities of nitrogen oxides, NO_2 and NO_3, which can be used by plants. How much is produced? Well, it is a difficult question to answer with certainty because the length of each lightning bolt, the temperature produced by each bolt, and the global frequency of lightning bolts are highly variable. However, one estimate suggests that about 15 million tonnes of oxides of nitrogen are produced by lightning each year. This is about one tenth of the global application of nitrogenous fertilisers each year, a significant fraction. NO_2 can be absorbed directly by leaves through stomata, the microscopic pores in leaves through which CO_2 (for photosynthesis) and water vapour (as transpiration) diffuse and contribute to the nitrogen metabolism of the plant. NO_2 can get "washed-out" of the atmosphere during rain to produce acid rain, which although harmful in excess, can benefit plant growth by increasing the amount of nitrogen that can be absorbed by plants roots. Thus, it is clear that lightning contributes substantially to the nitrogen budgets of global ecology.

Another global impact of lightning, however, might not be considered so benign….

LIGHTNING AND FIRE ECOLOGY

Lightning has the capacity to cause fires in natural landscapes, including savannas and boreal and tropical forests. Savannas, those open landscapes dominated by grasses with a sparse cover of trees, cover almost 25 % of the global terrestrial land surface. They can be found in Australia, Africa, north and south America, and Asia. Boreal forests (also known as taiga) are forests that grow in cold northern regions of Canada, Russia, Europe, and Scandinavia and are dominated by coniferous trees, including pines, fir, and spruces (Chapter 4). They cover about 12 million km^2 or about 24 % of the terrestrial land surface. Both savannas and boreal forests are subject to repeated and extensive burning. Savannas in northern Australia, for example, burn on average once every two or three years. Catastrophic fires in south Australia, such as the Black Saturday fires in Victoria (Feb – Mar 2009) burned about 450,000 ha (4,500 km^2) while the NSW fires in 1974/5 burned about 4.5 million ha. The Californian 2017 fire season saw about 9000 fires burn across the state, and about 5600 km^2 of land burned, and is considered the worst fire season on record. Even southern Europe (especially Spain and Italy) has suffered from extensive wildfires in the past decade (Chapter 12).

Fires damage property, infrastructure, tourism, and whole communities of people. Fires also have significant impacts on the ecology of woodlands and forest (see below). But does lightning really start many forest fires? It's a tough question to answer, but recent studies are showing just how they do contribute to fire ignition. In one study in Boreal forests of Canada, the average annual fire initiation rate from lightning was estimated to be 116 fires per year across a 91,000 km^2 forested area, that is, almost 1 fire per year per 1000 km^2 of forest. In the south-west of the USA, it was estimated that lightning caused 40 % of all wildfires reported between 1992 and 2013. Proportional contributions from lightning depend on the geography of the region, with lightning accounting for more than 98 % of reported wildfires in the more mountainous regions and only about 15 % in coastal and desert regions. However, globally, it is clear that lightning is responsible for a significant fraction of all wildfires in grasslands, savannas, woodlands, and forests.

A recent NASA-funded study concluded that lightning was the main cause of recent massive fire years in both Alaska and northern Canada (https://www.nasa.gov/feature/jpl/lightning-sparking-more-boreal-forest-fires). There was a record number of lightning-ignited fires in the Canadian Northwest Territories in 2014 and in Alaska in 2015. This study also concluded that lightning-induced fires in these regions have been increasing by 2 – 4 % per year since 1975, a result ascribed to increasing temperatures, which increase the number of storms but also increased plant growth (and hence fuel load) in these boreal regions.

Fires require about 16 % O_2 in the atmosphere (because fires consume oxygen). It wasn't until after the second oxygenation event (called the Neoproterozoic Oxidation Event; about 600 Mya) that fires could be sustained. The first Great Oxygenation Event started about 2.5 Gya, but it didn't lift O_2 above about 10 % of current levels, that is, to 2 % O_2. Because fires have been widespread and recurrent since the Neoproterozoic Oxidation Event, it is logical to hypothesise that landscapes that experience fire across millennia will evolve in response to this recurrent stressor. This raises the question: how have wildfires affected terrestrial ecology?

IMPACTS OF LIGHTNING AND WILDFIRES ON PLANT EVOLUTION AND BEHAVIOR, AND ATMOSPHERIC CHEMISTRY

Lightning can kill trees outright, even very large ones. The loss of a single, very large tree may appear to be of little consequence. However, this is incorrect. Very large, canopy dominant trees have multiple significant effects on the local forest. Light levels beneath a dense closed forest canopy can be as little as 2 – 10 % of that reaching the top of the canopy, and this limits the rate of photosynthesis and hence growth of seedlings, saplings, and understorey shrubs. The loss of a large canopy dominant tree opens the canopy, allowing more light to reach the forest floor. This sudden increase in light supply, across hundreds of square metres of the understorey, results in a race-to-the-top as seedlings and saplings, previously held back by the lack of light, race to fill the gap produced by the death of the large canopy dominant. The physiology and allocation of growth (increased allocation to numerous, small thick leaves instead of a few large thin leaves, for example), of these suppressed plants is altered in the arms-race to become the next canopy dominant. The death of the large dominant tree also releases, slowly, nutrients from the rotting root system and canopy leaf litter, which aids the growth of the surrounding sub-dominant trees that were once competing directly for these nutrients.

The impact of fire, caused by lightning or otherwise, tends to be more widespread, of course, than that caused by the loss of a single large tree. Hundreds, indeed hundreds of thousands, of square kilometres of grasslands, savannas, woodlands, and forests can burn during a single conflagration that lasts several weeks/months. Fire has been a recurrent feature of terrestrial landscapes ever since there was sufficient biomass available to burn AND sufficient O_2 in the atmosphere. Although O_2 levels started rapidly increasing in the atmosphere (because photosynthesis releases O_2) about 800 Mya (i.e., 0.8 Gya) but it wasn't until about 420 Mya that charcoal appears extensively in the geologic record, indicating that this was when there was both sufficient dry biomass and sufficient O_2 for that dry biomass to burn.

Fires are surprisingly varied. Fires in the early part of a dry season tend to be less intense (cooler) than fires late in the dry season when fuel has had longer to dry. Low stature grass fires tend to be shorter lived and cooler than fires that burn woody debris. Canopy fires can move extremely quickly through a forest. A low intensity grass fire might release up to about 500 kW of energy per metre of fire front and the flames might by less than 0.75 m in height. A moderate fire in a woodland might release 1000 – 3000 kW of energy per metre of fire front and flames might be 5 – 6 m high. An extreme fire storm can release more than 10,000 kW of energy per metre and flame height can be tens of metres, resulting in canopy fires (as well as stem burning).

Apart from the obvious death of the grass layer, understorey herbs and mid- and upper-strata trees, what other impacts do these fires have on the ecology of savannas, woodlands, and forests?

Many landscapes experience fire at a sufficient frequency for it to affect the evolution of a number of adaptive traits. Most of Australia, and large fractions of the remaining continents (excluding Antarctica), are fire prone. Consequently, many tree species protect their seeds from the heat of fire in thick, dense, woody fruits (fir cones, for example). Most importantly, these remain closed whilst mature until a fire has passed through (thus protecting the seed contained in the cone from extreme heat). Such species display *serotiny* – an adaptation whereby an environmental trigger (e.g., fire, drought) is required to initiate seed release. Thus, a fire is required to begin the release of seed to the ground. Many *Hakea* and *Banksia* species in Australia, and many coniferous species in the USA and Europe, retain seed in woody structures until fire passes. In some cases it is the heat that initiates seed release, but in some cases it is chemicals in the smoke that start the release of seed. Why exhibit such a strategy? One reason is to take advantage of the pulse of nutrients released into soil from ash, which contains a myriad of nutrients that become available to newly emergent seedlings when it rains. Second, the fire removes the understorey and often a small-to-large fraction of the upper canopy. This increases the amount of light available at ground level, supporting the rapid growth of the seedlings in the initial stages of the race for canopy dominance. In the months after a fire, there can be more than 200,000 seedlings per hectare (20 m^{-2}). After 5 years, this number declines to 10,000 – 20,000 per hectare and after 150 years this will be almost at an equilibrium density of about 50 – 100 trees per hectare. This massive change in tree density has a dramatic impact on catchment water yield in, for example, Mountain Ash catchments (water yield is the amount of water running out of the bottom of the catchment as stream and river flow). Water yield is maximal immediately after a fire because vegetation is no longer transpiring water to the atmosphere. As hundreds of thousands seedlings germinate and grow, their water use increases rapidly and catchment water yield declines for 20 – 40 years. As self-thinning increases and tree density decreases, water yield gradually increases and reaches a new equilibrium at some point around 75 – 140 years after the fire.

Most eucalypts, after a reasonably hot fire, look dead, with no leafy canopy and charred bark. However, epicormic buds lie dormant under a thick protective bark and sprout, often from tree stems, soon after a fire. The presence of epicormic buds is a widespread adaptation to coping with fire and this trait has been used by foresters for more than 2000 years in pollarding and coppicing woodlands to maintain wood production despite harvesting trees for construction purposes. An additional adaptation present in eucalypts is that of a *lignotuber*, a large dense underground storage organ, from which multiple stems can regenerate if the entire above-ground trunk is killed by fire. The lignotuber stores water, nutrients and numerous epicormic buds that are activated if the trunk of the tree is killed.

FORMATION OF OZONE AND NOx AND THE OZONE HOLE

If you have ever been close to a ground strike of a lightning bolt, as I was in northern Australia 30 years ago, when a 2 m tall sapling was vaporised by a lightning bolt about 100 m from where I was standing, you may have noticed a distinctive smell similar to that in a room full of photocopiers. That is the smell of ozone (O_3). The word ozone comes from the Greek word meaning smell. High levels of ozone are toxic to humans, and to germs.

At ground-level, ozone is toxic to life, causing damage to lungs, plant leaves, and most other forms of life because it is a very strong oxidising agent and rapidly damages cell membranes and proteins, leading to cell death. In research conducted in Europe, I demonstrated that ozone reduces frost hardiness of conifers (Chapter 4). Ozone has been used as a disinfectant commercially for treating potable water by dissolving it in water to produce a broad-spectrum biocide for more than 110 years. It is active against bacteria, viruses, and cysts of parasites.

Ground-level O_3 is created by chemical reactions between oxides of nitrogen (NOx) and volatile organic compounds (VOC) in the presence of sunlight and warm temperatures. Many VOCs, including limonene and isoprene, are released by plants, while NOx is released by internal combustion engines and power stations due to the high temperatures used. Lightning also produces NOx (and O_3) at each bolt of lightning. Indeed, about 10 % of NOx at ground-level is produced by lightning and this NOx can result in additional ozone being created naturally by the interaction of NOx and VOCs. Unhealthy levels of O_3 (unhealthy to plants and humans) are recorded across many days in summer in cities and towns globally, although increased regulation of air quality is reducing the prevalence of damaging levels of ozone.

In the early 2000s, researchers funded by NASA (and others) demonstrated that lightning in the lower troposphere (about 5 to 13 km above ground-level) above the USA was producing more O_3 and NOx than was being generated by human activities. Prior to this research, quantification of rates of lightning-induced tropospheric O_3 had not been made. Both the production of O_3 directly, and also the production of O_3 through the presence and transformation of NOx, are generating significant levels of O_3 in the lower atmosphere. This is undoubtedly also true for the rest of the world.

In the upper atmosphere, O_3 can be generated through the action of UV radiation (photons in the waveband 120 – 400 nm) on O_2. The O_2 disassociates to form 2 oxygen atoms and one of these can interact with O_2 to form O_3. This occurs predominantly between 10 – 50 km above the Earth's surface in the stratosphere, producing the "ozone layer" although even in the "ozone layer" the concentration of O_3 is low (about 2 – 8 parts per million). Although low, this level is sufficient of prevent most (but not all) of solar UV reaching the ground and damaging the DNA of living organisms (O_3 absorbs UV radiation in the waveband 100 – 315 nm). During the 1960's and 1970s, O_3 concentrations in the stratosphere exceeded about 220 Dobson units (1 Dobson unit is equivalent to 2.6867×10^{20} molecules per square metre). However, during the 1980s, O_3 concentrations began to decline significantly in the stratosphere above the south polar region during the spring of each year – the so-called "ozone-hole". Concentrations of O_3 declined to less than 100 Dobson units. The cause of this decline was the increasing concentration of chlorofluorocarbons. The global banning of the use of these chemicals in refrigeration systems and aerosol cans (as a propellant) in 1987 has resulted in a gradual increase in stratospheric O_3 levels and a reduction in the size and extent of the O_3 hole. We might expect a full recovery sometime in the middle of the 21st century.

BALL LIGHTNING, AIRPLANES, AND ST. ELMO'S FIRE

The existence of ball lightning has been confirmed through public observation for almost 1000 years. An English monk, Gervase of Canterbury, may be the first person to write an account of ball lightning in June 1195, although mythological accounts predate this by millennia. Personal accounts of ball lightning, often from "people of repute", including churchgoers in 1638, sailors on the British sloop Catherine and Mary in 1726, an Admiral of the British fleet in 1749, a Russian Professor in 1753, and multiple passengers and crew (including military pilots) on aircraft in 1960s – 2010s, and even submariners, have been made in almost every decade over the past several hundred years. Reviews of personal accounts of ball lightning have identified many thousands of cases.

So, what is ball lightning? Visual descriptions vary, but some consistencies are apparent. They can be small (< 1-cm diameter), medium (< 1 m) or large (> 1 m), spherical luminous or fire-like (fiery) objects that move through the air at slow-to-fast speeds (walking pace-to-running speeds), in relatively straight or curved trajectories moving upwards, downwards, or both. They may persist for a few seconds to several tens of seconds. They frequently explode at the end of their life, sometimes leaving a distinct sulphurous smell. Interactions with physical objects, including walls, windows, and people are relatively common and can result in the death of individuals. Accounts of ball lightning passing through walls without affecting the walls have been made and ingress and egress of a single airplane has been recorded, without any apparent damage to the aircraft. Of course, these are visual descriptions of the phenomenon, rather than an answer to the question: what is ball lightning?

Ball lightning is entirely unpredictable. We can't know where or when it will occur, hence our reliance on the billions of human eyes to provide all the evidence for ball lightning over the past millennia. But all of that changed in 2012 when a research group at the Northwest Normal University in Lanzhou, China was making measurements of run-of-the-mill cloud-to-ground lightning on the Tibetan Plateau (a vast plateau located in Tibet, northern and western China, Pakistan, India, Nepal, Tajikistan, and Kyrgyzstan). By pure chance, a ball of lightning occurred within the study site, which was scattered with ultra-high-speed cameras, spectrometers, and video cameras. Perhaps the most exciting data recorded were the emission lines of the phenomenon (Fig. 3). This spectrum clearly identified emissions corresponding to silicon, calcium, iron, nitrogen, and oxygen. This was in marked contrast to the spectrum of the lightning bolt that immediately preceded the formation of the ball lightning, which, as is normal, was dominated by the emission lines associated with ionised nitrogen. The ball lightning was about 5 m in diameter and travelled horizontally at a speed of about 8.6 m s^{-1} and lasted less than 2 s.

Can these data help explain the origin of ball lightning? One of the theories accounting for the origin of ball lightning is that the magnetic field associated with lightning strikes induces hallucinations in the visual cortex of the brain by stimulating neurons to fire in a pattern that can resemble that of ball lightning.

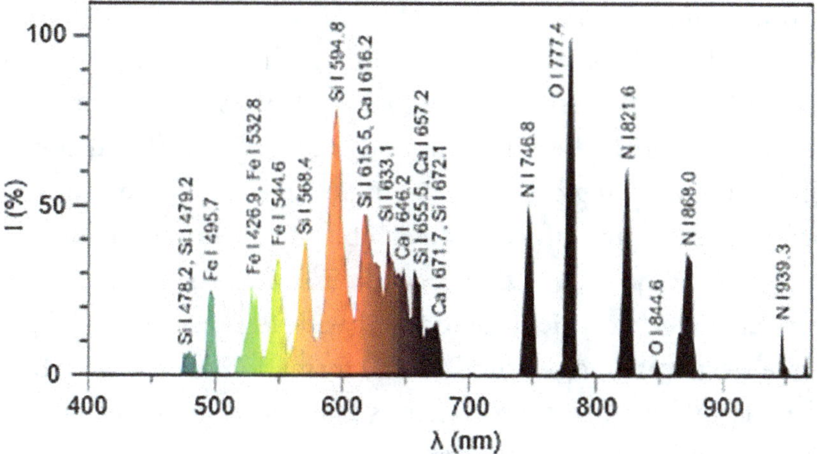

Figure 3: Emission spectrum of a ball lightning and emission peaks of silicon, iron, calcium, nitrogen, and oxygen. Vertical axis: Intensity (%), horizontal axis: wavelength (nm). Data from Cen, Jianyong; Yuan, Ping; Xue, Simin (2014). Observation of the Optical and Spectral Characteristics of Ball Lightning, Phys. Rev. Lett. 112,doi:10.1103/ PhysRevLett.112.035001. From: https://en.wikipedia.org/wiki/Ball_lightning#/media/ File:Ball_lightning_spectrum.svg. This image is made available under the Creative Commons CC0 1.0 Universal Public Domain Dedication. Author: Olli Niemitalo

An alternative hypothesis is that ball lightning occurs when normal lightning hits the ground and vapourises a small patch of soil. If this were true, the spectrum of the ball lightning should resemble the spectrum of vapourised soil, which tend to contain a lot of silicon, calcium, iron, nitrogen, and oxygen (the "soil clod hypothesis"). The Tibetan Plateau spectrum, the only one-of-its-kind, certainly supports this hypothesis. How ball lightning can appear up in aircraft, or in submarines, however, remains a mystery!

St. Elmo's fire

Ball lightning is not the same as St. Elmo's fire. References to St. Elmo's fire extend back to the Roman empire and was noted by Charles Darwin when he was on The Beagle. Shakespeare refers to them in The Tempest, and the 1956 movie Moby Dick refers to them. Even the 2017 Eurovision song contest entry from San Marino refers to St. Elmo's Fire.

St. Elmo's fire consists of a luminous plasma generated by a corona discharge from the end of a rod-like structure (mast of a yacht, church steeple, flagpole)

within a strong atmospheric electric field. (A corona discharge is an electrical discharge caused by the ionization of air surrounding a conductor (for example, a metal flagpole that has been hit by lightning)) carrying a high voltage. It can be seen in a local region where the air has undergone electrical breakdown and become conductive, allowing charge to continuously leak off the conductor into the air. The surface edges of aircraft can also exhibit the phenomenon of St. Elmo's fire. A key feature of St. Elmo's fire is that it can be readily produced in the lab.

The very large electrical fields associated with the discharge of lightning provides the large electrical field required for the formation of St. Elmo's fire. The large electrical field is sufficient to strip electrons from gaseous molecules in air. Molecules of nitrogen and oxygen in the atmosphere fluoresce in the blue/violet part of the spectrum. Neon lights contain the gas neon and therefore produce orange light, while hydrogen emits red, and helium produces yellow light. Such fluorescent lights contain a plasma (ionised gases) within them to generate the light emitted that we see. Plasma TVs use small cells, each containing a plasma that is manipulated using electric fields to produce the image we see.

In conclusion, lightning is a complex phenomenon that relies on the friction that occurs within clouds to generate positive and negative charges on ice crystals within clouds. The lower portion of the cloud tends to be negatively charged (negatively charged graupel falls downwards under gravity) and the upper portion positively charged as the lighter and positively charged ice crystals float upwards.

The negative charge on the lower surface of the cloud induces an equal and opposite (positive) charge to accumulate on the ground immediately below the cloud. The air between the base of the cloud and the ground surface is an electrical insulator and very large voltage differences accumulate between the cloud base and ground surface. Once this voltage is large enough, it is able to "jump" from cloud-to-ground; lightning bolts can exhibit voltages of tens-of-millions to 300 million volts.

Lightning affects the chemistry of the lower atmosphere by producing significant quantities of NOx and O_3, which of themselves affect the ecology of terrestrial environments. Lightning-induced fires affect the evolution and ecology of landscapes, especially savannas, woodlands, and forests, globally.

Ball lightning and St. Elmo's fire are two (separate) phenomena with a long history of observation and mythology. While St. Elmo's fire is a well understood and reproducible phenomenon, explanations of ball lightning remain elusive, although the rare spectral analyses provided from a single, accidental observation in the Tibetan Plateau, supports the so-called "soil-clod" hypothesis.

THE END

Chapter 10

As indicated in the Preface, this was the first science-based question (well, the first part about the sky) that I recall asking my Dad about 55 years ago. He didn't have an answer, but we spent several hours looking through a set of Encyclopedias he and Mum had recently bought, to find an answer. It is undoubtedly his fault that I became a scientist.

In this chapter, I start with a consideration of the wavelengths (colours) of light emitted by our Sun, before discussing the absorption and scattering of photons by gas molecules and dust in the atmosphere. The reason for the sky being blue on Earth, but black when viewed from the moon is all to do about scattering of photons by an atmosphere. When examining the question about why grass is green, some weird similarities between the structure of haemoglobin in Human blood, chlorophyll in plant leaves, foetal haemoglobin in the blood of an unborn baby, and the nodules of roots of legumes (peas and beans, for example) are uncovered. We also discover that chlorophyll can be bright red, despite usually looking green. I then spend time talking about how the field of remote sensing from satellites uses this fact to measure the greenness of terrestrial landscapes and show some significant increases in terrestrial landscape "greenness" globally, and the importance of this greening. Finally, we see how global decreases in sea levels recorded in 2010/2011 were detected by a system measuring the 'breathing of landscapes' located in the middle of Australia.

THE TEMPERATURE OF THE SUN AND WAVELENGTHS OF PHOTONS EMITTED

The sun is hot. The temperature of the surface is almost 6000 °C. The temperature of an object determines the colour (the wavelength) of the light it emits. An iron bar heated to 'red hot" is about 600 – 700 °C. At that temperature it emits a lot of light (photons) with a wavelength of around 650 – 700 nm (nm means nanometre, or 1 billionth of a metre or 1 millionth of a millimetre). Photons with a wavelength of 650 – 700 nm are red. The hot iron bar emits photons in the infra-red waveband too, but we can't see those. As the temperature of an object increases, more and more photons are emitted at shorter and shorter wavelengths. Remember blue is a shorter wavelength (around 450 nm) than red (around 680 nm). At about 1200°C the iron bar now emits far more photons at lower wavelengths, in addition to the longer red wavelengths. This *mixture* of wavelengths makes the iron bar look "white hot".

The sun is much hotter than the 1200 °C iron bar. It emits a flood of radiation, at wavelengths between 250 and 2500 nm (Fig. 1). The broad peak of emissions of solar radiation is about 400 – 700 nm, which just happens to be highly important

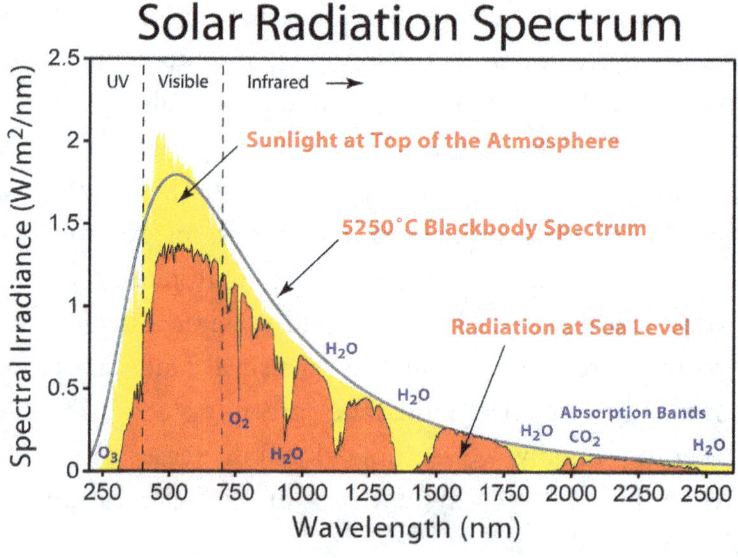

Figure 1: The solar radiation spectrum, as measured at the top of the atmosphere, and at sea-level, demonstrating the absorption of different wavelengths of light by the atmosphere. This figure was prepared by Robert A. Rohde as part of the Global Warming Art project. Permission is granted to copy, distribute and/or modify this document under the terms of the GNU Free Documentation License, Version 1.2

for two unrelated reasons. First, the waveband 400 – 700 nm is the visible spectrum, this is what our eyes see. Second, this waveband is also called Photosynthetically Active Radiation (PAR), the waveband that plants use to drive photosynthesis, which produces, directly and indirectly, all the food we eat...... but more of that later.

The top of the atmosphere is bombarded with photons of wavelengths between 250 and 2500 nm. Pink Floyd showed us in 1973 that white light, the light coming from the sun, is composed of all the colours of the rainbow – red, orange, yellow, green, blue, indigo, and violet. Of course, Pink Floyd were not the first to show this effect – Isaac Newton did this in 1672! The point is, solar radiation (sunlight) is composed of a wide range of wavelengths, and our eyes can differentiate the visible spectrum into the seven colours of the rainbow. What is rarely understood is that it is possible to recombine these seven colours to produce white light – Newton did this is 1672 too!

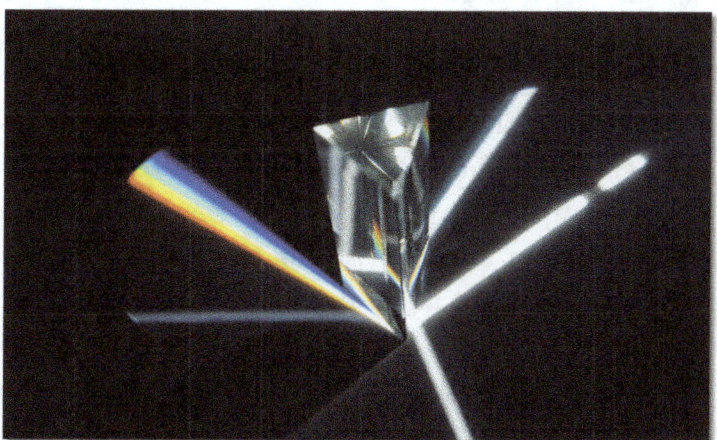

Figure 2: Refraction of white light as it passes through a prism, redolent of the iconic Pink Floyd Dark Side of the Moon album cover. This image is made available under the Creative Commons CC0 1.0 Universal Public Domain Dedication. From https://commons. wikimedia.org/wiki/File:Dispersive_prism.png

SO WHY IS THE SKY BLUE, NOT WHITE?

Scattering is the answer. More specifically, Rayleigh scattering, named after John William Strut, 3rd Baron of Rayleigh, who discovered this type of scattering in the 19th century in England. What does "scattering" mean in this context? When a white snooker ball (the colour of the ball is irrelevant) makes a glancing hit on a red snooker ball (the colour is of the ball is irrelevant), the path of the white ball is altered and could be said to have been scattered by its interaction with the red snooker ball. Scattering is simply the change in direction of the photon after it interacts with dust or gas molecules.

Water molecules have a diameter of about 0.2 nm, CO_2 molecules are about 0.232 nm across, and molecular nitrogen (N_2) is about 0.3 nm across. These are not too different (relative to the size of a brick, let's say) from the wavelength of visible radiation (400 – 700 nm). Because the wavelength of blue light (400 nm) is closer to that of the diameter of gas molecules in air, they are scattered about 10-fold more than the red photons, which have a wavelength of around 680 nm. When the sun is overhead (solar noon, which may or may not correspond to local time-zone midday), the scattering of blue photons occurs at all points in the atmosphere and a large fraction of the scattered photons are directed down towards Earth. This increases the fraction of blue photons that arrive at our eyeballs (the blue photons coming directly to us from the sun plus the blue photons that have been scattered downwards). Consequently, all the sky appears blue when the sun is overhead. Scattering of red and yellow photons is minimal so when we look at the sky away from the sun, there is no enhancement of the fraction of red/yellow photons arriving at our eyeballs, further enhancing the blueness of the sky. This is true for most of the day, that is, between one hour after sunrise and up to one hour before sunset. Put another way, the larger degree of scattering of blue photons (relative to red and yellow photons) increases the proportion of blue photons we see and so the sky looks blue. Looking more closely towards the region of the sky containing the sun disc itself, we see the sun and region of sky close to it is more yellow than blue. This is because the red and yellow photons are least scattered and what we are receiving from this region of sky is the unscattered yellow and red photons coming directly to us from the sun. In addition, there is a significant depletion of blue photons that have been scattered by the oxygen, nitrogen, and water molecules in the atmosphere in directions away from where we are standing.

But why isn't the sky violet since violet photons have an even short wavelength than blue and so are scattered more than blue photons? The human eye has three types of colour sensitive cells in the retina (cone cells). The peak sensitivities of these cones are (a) 420 – 440 nm, (b) 534 – 555 nm, and (c) 564 – 580 nm. Because none of the peak sensitivities are in the violet region, but one is in the blue region, we see the blue photons much better than we see the violet.

WHY ARE SUNSETS RED?

When the sun is overhead, photons travel through about 480 km of atmosphere (although most of the gases are contained within the lowest 20 km, close to Earth). When the sun is setting, photons travel through much more of the atmosphere and more of their path is within the 20 km of Earth where the majority of atmospheric gases and particulate matter are found (Fig. 3). Consequently, most of the blue photons are scattered away from our direct line of sight, but the red and yellow photons travel through this extra thickness of atmosphere without being scattered, giving sunsets (and sunrises) an enhanced red/orange colour.

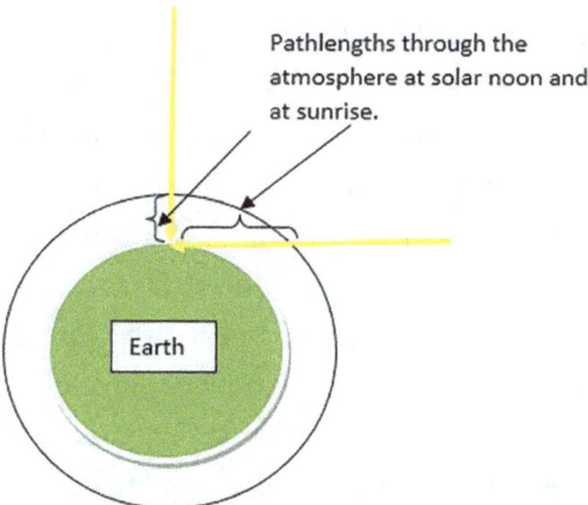

Figure 3: A schematic diagram demonstrating how the pathlength of a photon through the atmosphere is longer in the morning and evening than at midday. The yellow arrows represent incoming solar radiation.

WHY IS THE SKY BLACK WHEN VIEWED FROM THE MOON?

Because photons from the sun going across the sky to somewhere else (that is, not into our eyeballs) are invisible, we don't detect them. In a vacuum, a beam of light from a torch is invisible unless it is directly pointing into our eyes. On Earth, we only see photons that have been scattered by the atmosphere and clouds, or that arrive directly from the sun, into our eyeballs. When we look at the sky away from the Sun, on Earth, the only photons we see are those that have been scattered by the atmosphere and clouds. If we were on the Moon's surface, when we look at the sky *away from the Sun and away from Earth,* the sky is black because there is no atmosphere on the moon, so there is no scattering of photons and so the sky is black. (When we look at Earth from the Moon, we see red, orange, yellow, green and blue photons arriving directly to our eyeballs after being reflected from the Earth's surface.)

Who would have thought about that, apart from my brother, a physicist?

BLUE LIGHT AND HUMAN HEALTH

If we assume that *Homo sapiens* is about 300,000 years old, then for more than 99 % of our evolutionary history, we have relied on natural sunlight, with some supplementation from firelight. The sky tends, for a significant fraction of the day, to be blue. The sea also tends to be blue, rather than red, or yellow, because water absorbs red and yellow wavelengths more effectively than blue wavelengths, and so the increase in the proportion of blue photons reflected to us makes the sea look blue. It is probably no accident that humans feel a sense of relaxation during the day when they experience blue skies and blue seas. Indeed, exposure to blue light can increase alertness, help memory, and brain function, and elevate mood. It is important in regulating our natural wake and sleep cycle (circadian rhythm). Blue light may also be effective in the local treatment of eczema and psoriasis. However, a key point to note here is that some of these benefits are *only apparent following exposure during the daytime.* Exposure to blue light at night, even very low levels, for example the amount emitted from digital devices (the screen and the blue light indicating the "on" status of such devices), for prolonged periods, has been associated with disrupted sleep patterns because it affects the production of melatonin. Interruption of the circadian system plays a role in the development of type 2 diabetes, cardiovascular disease, cancer, sleep disorders, and cognitive dysfunctions. Getting blue-light filters for computer screens and digital devices, particularly after sundown, is probably a smart move!

WHY IS GRASS GREEN?

Most healthy leaves are green. Why is this? The simple answer is because they contain a lot of chlorophyll. There are many types of chlorophyll, innovatively called chlorophyll *a*, chlorophyll *b*, and chlorophylls *c1*, *c2*, *d* and *f*. Chlorophyll *a* is found universally across plants, algae, and cyanobacteria (bacteria-like organisms that lack a nucleus and are capable of photosynthesis). Chlorophylls *c1* and *c2* are found in algae while chlorophylls *d* and *f* are found in cyanobacteria. Strangely, chlorophyll is a registered food colorant; E140. Chlorophyll is used to make some pastas and the drink absinthe, green.

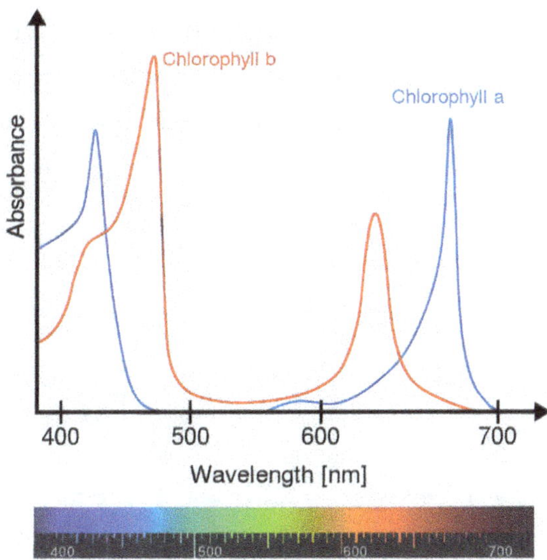

Figure 4: The absorption spectrum of chlorophylls *a* and *b*. This image is licensed under the Creative Commons Attribution-Share Alike 3.0 Unported license. Author: Daniele Pugliesi. From https://commons.wikimedia.org/wiki/File:Chlorophyll_ab_spectra-en.svg

Chlorophyll contains five rings made up of C and N atoms and these rings are all in the same plane, surrounding a central magnesium ion (Mg^{2+}; Fig. 5). There are many alternating double and single bonds between adjacent atoms (of C or N). This molecular structure means that electrons are "shared" or "smeared" along the ring and this means that they absorb red and blue photons of light really efficiently, but they don't absorb green photons of light, which are therefore reflected or transmitted. Our eyes see these green photons so leaves look green to us; all the red and blue photons have been absorbed by the chlorophyll molecules that sit in chloroplasts of leaves of plants. The physical basis of "absorb" is described in a section below caller "Photochemistry". See Figure 9 too.

AN INTERESTING ASIDE: HAEMOGLOBIN AND CHLOROPHYLL

Chlorophyll is central to photosynthetic carbon fixation by cyanobacteria (also called Cyanophyta; a phylum of bacteria), algae, and higher plants. Haemoglobin is central to carrying oxygen in the blood of almost all vertebrates.

Chlorophyll possesses a central magnesium ion, Mg^{2+}, while haemoglobin possesses a central Fe^{2+}. These central ions are each surrounded by four pyrrole rings (5-member rings containing 4 C atoms and an N atom, with two double bonds present in the ring; Fig. 4). The structural similarities are startling, and yet the evolution of haemoglobin occurred much later than that of chlorophyll (more than 1 billion years later).

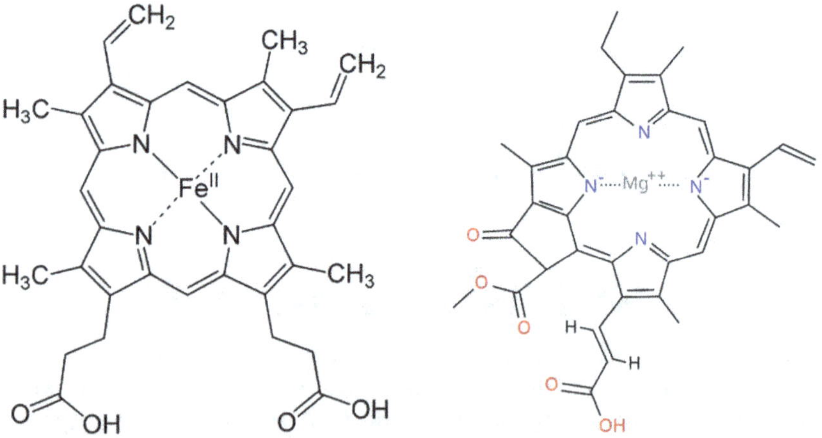

Figure 5: The molecular structure of Heme B (left image; American spelling; Haem B, British spelling) and chlorophyll *c1* are remarkably similar. Haem B is found in haemoglobin and myoglobin in mammals. Chlorophyll is found in cyanobacteria, algae, and plants. Author of the Haem image: Yikrazuul; Public domain licenced. From: https:// en.wikipedia.org/wiki/Heme Chlrorophyll image by David Richfield, licensed for the public domain. From: https://en.wikipedia.org/wiki/Chlorophyll

But the weirdness doesn't stop here. The haemoglobin in a human foetus is called foetal haemoglobin. Its function is to "pull" oxygen from the mother's blood into the foetus and to do that it must have a larger affinity for oxygen than the mother's haemoglobin. If this wasn't the case, the foetus would be unable to extract oxygen from the mother's blood and would not develop. Foetal haemoglobin is structurally very similar to myoglobin, a haem protein found in vertebrates responsible for supplying oxygen to muscles. Nitrogen-fixing plants (for example, peas, runner beans, soybeans) have bacteria growing in nodules on their roots. If you split these nodules with your fingernail or a knife, they are pink/red inside. They are pink/red because they contain leghaemoglobin, an Fe-containing molecule that attaches to oxygen diffusing into the nodule. This is necessary because the nitrogen-fixing enzyme (nitrogenase) is very sensitive to the presence of oxygen – it rapidly loses enzymatic activity when exposed to too much oxygen. Leghaemoglobin is structurally similar to myoglobin and foetal haemoglobin. Indeed, leghaemoglobin (in legumes) and animal haemoglobin evolved from a common ancestor.

Figure 6:   The nodules on roots of legumes are pink because they contain leghaemoglobin, which is extremely similar in structure and colour to the haemoglobin in human (and other animal) blood.

ANOTHER WEIRD THING: CHLOROPHYLL IS GREEN, BUT ALSO RED

The upper picture in Figure 7 shows a flask containing chlorophyll extracted from a leaf, illuminated with white light. The chlorophyll is, as expected, green. However, the same extract, illuminated with UV, shows the chlorophyll to be red.

Figure 7:   A solution of chlorophyll illuminated with white light to simulate solar radiation, looks green, but when illuminated with UV, it glows red (lower picture). This image is licensed under the Creative Commons Attribution-Share Alike 3.0 Unported license. Author: Marie Franzen. From: https://en.wikipedia.org/wiki/Chlorophyll_fluorescence

Similarly, the upper panel in Figure 8 shows a microscopic image of the surface of a moss leaf, illuminated with white light. The green colour within each cell is the chlorophyll. The lower image was recorded using a fluorescence microscope and the chlorophyll is now red.

What is going on here? To have a go at answering this, we need to know a little about how the energy in photons is captured by chlorophyll during the process of photosynthesis.

Globally, photosynthesis exceeds all other chemical synthetic processes. Approximately 100 billion tonnes (10^{11} tonnes) of C are fixed through photosynthesis each year and in the process, 265 billion tonnes of O_2 are released, for which we aerobes are eternally grateful. About half of the C fixed by plants is subsequently lost through respiration of those plants, but the rest accumulates as biomass, which in turn supports 99.999 % of all life on Earth.

Photosynthesis consists of three linked and simultaneously occurring processes:

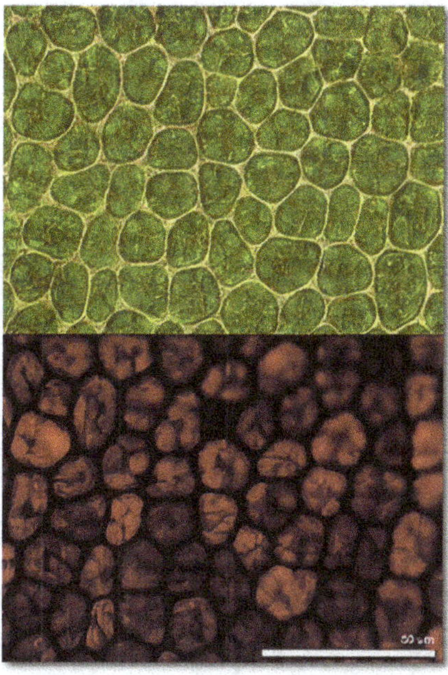

Figure 8: Microscopic images of a moss leaf from *Plagiomnium undulatum*. Bright-field microscopy at the top and fluorescence microscopy at the bottom. The red fluorescence is from chlorophyll in chloroplasts. This file is licensed under the Creative Commons Attribution-Share Alike 4.0 International license. Author: Dietzel65. From https://en.wikipedia.org/wiki/Chlorophyll_fluorescence

(1) Photochemistry – the absorption of light by pigments in chloroplasts

(2) Electron transport – the transport of electrons to produce ATP and NADPH

(3) Biochemistry – the fixation of CO_2 and production of the first stable products of photosynthesis.

We are only concerned in this chapter with the first process: photochemistry.

PHOTOCHEMISTRY

Photochemistry is the trapping of photons of light (and hence the capture of their energy) in leaf pigments and this occurs in chloroplasts, those green sub-cellular organelles that make leaves look green. Specifically, we are concerned with electron absorption by molecules of chlorophyll in chloroplasts. A photon of light can be absorbed by a chlorophyll molecule located in the internal membrane of chloroplasts. Chlorophyll *a* is the dominant form of the two major types of chlorophylls in higher plants (chlorophyll *a* and chlorophyll *b*). Chlorophyll *a* and *b* have absorption peaks of 430 nm and 664 nm and 460 nm and 647 nm respectively, so chlorophylls *a* and *b* have a maximum rate of absorbance at these wavelengths. To understand how chlorophyl molecules capture the energy of a photon, we need to know a little about atomic/molecular structure and electrons. In the following brief discussion I refer to electrons as though they are particles occupying a single point in space. This is one way to consider electrons. However, electrons also possess the properties of a wave. The same can be said of photons, which simultaneously have the properties of particles, behaving like billiard balls during collisions, but also the properties of waves, being diffracted like waves in the sea or sound waves. This is irrelevant for our discussion below, but I did not want you to think electrons only exist as particles.

In a hydrogen atom (H), there is one electron, and this electron is located in a shell, or orbit, around the nucleus. This single electron occupies the orbit called the K shell, or 1 shell (I have no idea why it isn't called the 1st shell), the orbit closest to the nucleus. The maximum number of electrons that can occur in the 1 shell is two. However, atomic H is unstable, but H_2 (the usual form of hydrogen gas found in the atmosphere is relatively stable (unless you put a match to it), compared to H, as it contains two electrons. The 1 shell now contains the maximum number of electrons that can occur in the 1 shell, making molecular hydrogen (H_2) more stable than atomic hydrogen (H) – full outer electron shells make for a more stable atom.

The next shell/orbit out from the nucleus is the 2 shell, or L shell. This can contain a maximum of eight electrons. The next shall/orbit out is the 3 shell of M shell. This can contain a maximum of 18 electrons. More orbits are added as you move through the periodic table of elements.

A key point now is to define the *ground* and *excited states* of an atom. The ground state of an atom is the configuration of electron distributions across orbits that produces the lowest electronic energy (the sum of all energies of all electrons in that atom). The energy of an electron is determined by the orbital it is sitting in. So, the ground state is the lowest energy content summed for all electrons in the atom. The excited state arises when energy (for example, in the form of heat, or a photon hitting the atom) is absorbed by an atom and an electron moves from its ground state to an excited state. The amouunt of energy coming in (for example, the energy associated with a specific wavelength of light) must match the energy required to move the electron from the ground state to the excited state. This is a tightly defined amount of energy, just as the distance from one rung of a ladder to the next rung up is a specific and constant distance (for a particular ladder). If you try to climb a ladder using a fraction of the distance between rungs, let's say, half, or ¾ of the distance between rungs, you won't be able to climb the ladder. Similarly, if you try climbing 1.25 or 1.5 times the distance between rungs, you won't succeed in climbing the ladder. The distance between rungs is analogous to the amount of energy required to move an electron from the ground state to an excited state.

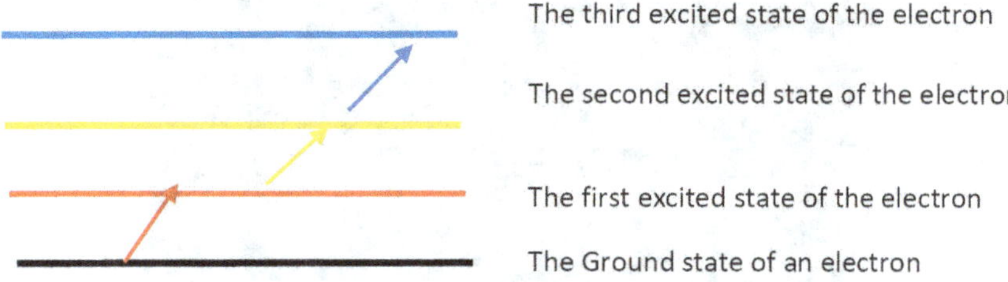

Figure 9: A crude schematic representation of the transition of an atom or an electron from the ground state to an excited state, with three levels of excitation represented. The three colours illustrate the three different amounts of energy that are required to go from the first to the second and third excited states.

An excited electron is very unstable, and the electron generally returns to the ground state very quickly, a process called decay. The decay of the atom (the return of the electron to a ground state) results in the release of energy in the form of a photon because the energy associated with the excited state (that is, the energy absorbed by the atom) must go somewhere. For chlorophyll, the release of energy during decay is in the form of red photons and the emission is called fluorescence. This is why leaves and chlorophyll illuminated with UV light (photons containing more energy per photon than visible light because UV has a shorter wavelength than visible light) causes the leaves/chlorophyll solution to fluoresce red. Fluorescence has many practical uses, including cathode-ray tubes (the old type of heavy and structurally deep TVs), medicine, spectroscopy, and washing powders (to give clothes that whiter-than-white, blue glow after a wash). Many species of fish, corals, squid, jellyfish, amphibians, butterflies, parrots, spiders, many types of rock/minerals, and all plants containing chlorophyll (most parasitic plants do not contain chlorophyll) are fluorescent (Fig. 10).

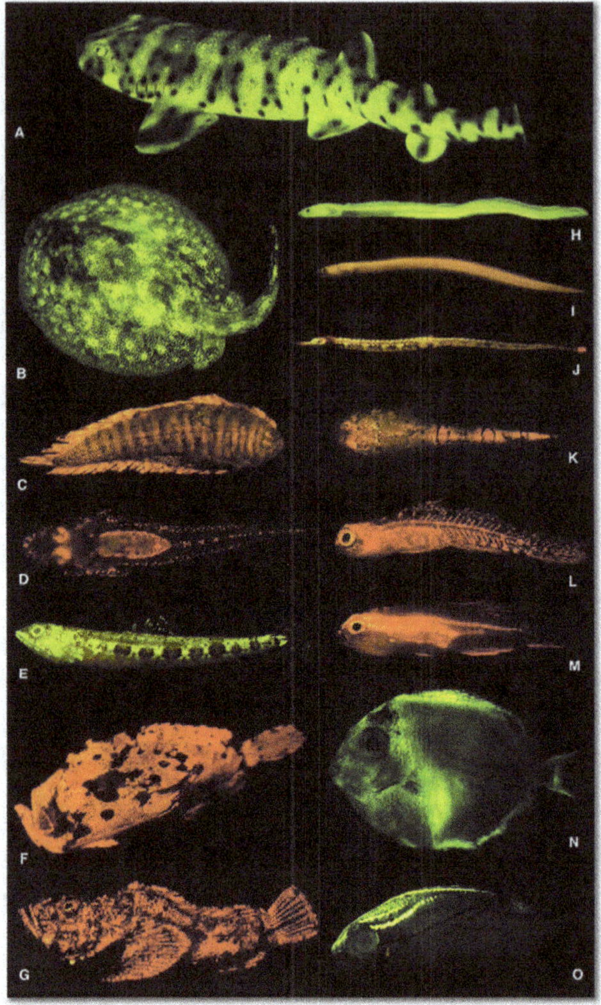

Figure 10: Many species of fish are fluorescent. This image is licensed under the This file is licensed under the Creative Commons Attribution 4.0 International license. Authors: Sparks, J. S.; Schelly, R. C.; Smith, W. L.; Davis, M. P.; Tchernov, D.; Pieribone, V. A.; Gruber, D. F. From: https://en.wikipedia.org/wiki/Fluorescence

Chlorophyll is a large molecule, not an atom. Indeed, it contains 137 atoms ($C_{55}H_{72}O_5N_4Mg$). While the picture is therefore more complex, with electrons shared and "smeared" across multiple atoms and the molecular orbitals of each molecule differing from other molecules with different compositions, the idea of excitation, decay, absorption of the energy of photons, and emissions via fluorescence are the same as those for atoms. Thus, chlorophyll absorbs blue and red wavelengths preferentially, and fluoresce by the emission of red photons.

MEASURING EARTH'S SURFACE FROM SPACE

In the 16th to 19th centuries, the only way to investigate surface features of Earth (for example, mountains, glaciers, rivers, forests, savannas) or the ocean surface (for example, sea surface temperature, algal biomass) was to place explorers in/on those features. Because there were so few explorers traversing the 510,000,000 km^2 surface of Earth, it took a long time to get even a rudimentary and spatially incomplete picture of the Earth's surface and its features.

That all changed when satellites were installed in space with sensors looking down at the surface of Earth. The science of Remote Sensing is the science of collecting information about Earth's surface without direct physical contact. Solar radiation (which spans wavelengths from 250 to 2500 nm; Fig. 1) travels through Earth's atmosphere, getting scattered and differentially absorbed by various components of the atmosphere, including CO_2, O_2, O_3 and H_2O. Earth reflects a variable fraction of incoming solar radiation and sensors on satellites detect this reflected radiation and quantify it. Different surfaces (for example, snow, oceans, rainforest canopy) preferentially reflect different wavelengths. Ice, being white, strongly reflects all the visible wavelengths (from about 380 nm to about 740 nm), but a rainforest canopy absorbs the red and blue wavelengths (and uses the energy contained in these photons in the process of photosynthesis to produce sugars and the universal chemical currency of energy, ATP (adenosine triphosphate; see Chapter 7) and reflects green wavelengths. It is the difference in reflectance properties of different surfaces that allows remote sensing scientists to infer the nature of the surface doing the reflecting (Fig. 11). In addition to reflection, all objects on the Earth's surface, including soils, rocks, oceans, vegetation, people, and houses, *emit* thermal and microwave (1 mm to 1 m wavelengths) radiation. The peak wavelength of thermal radiation is determined by the temperature of the object. Satellites have sensors to detect visible, UV, and thermal radiation reflected or emitted from Earth and these signals are beamed down to receiving stations on Earth and used to provide information about the Earth's surface. They also measure chlorophyll fluorescence from marine algae, crop fields, savannas, woodlands, and forests (see below).

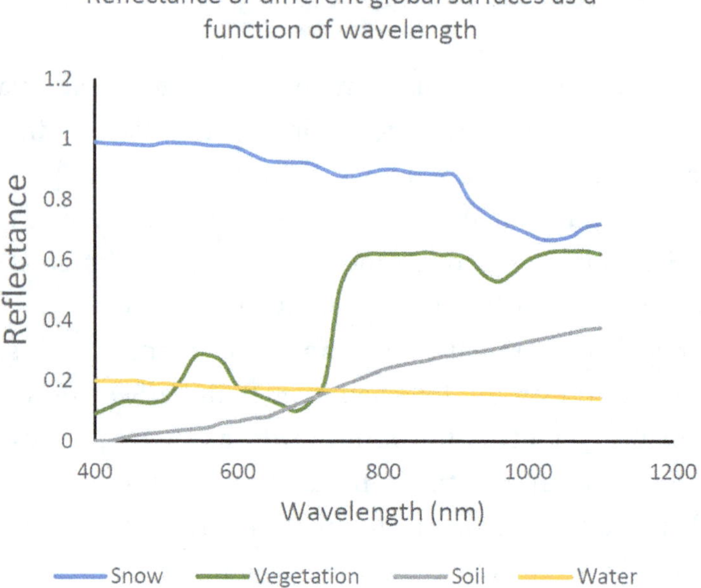

Figure 11: Reflectance of four classes of land-cover as a function of wavelength.

The range of information that can now be determined by remote sensing is phenomenal and includes (but is not limited to):

- The Earth's surface temperature (land and water)
- Areal extent of glaciers across the world
- Chlorophyll concentration of oceans
- Elevation
- Quantification of vegetation water stress (insufficient availability of water to plants, including forests, grasslands, and crops)
- Vegetation productivity, and changes in productivity at seasonal, annual, and decadal timescales
- Vegetation distribution and changes in distribution across time
- Leaf area index (LAI; a measure of how many square metres of leaf area present above any given square metre of land; a rainforest may have an LAI of 5 or more, a savanna may have an LAI of less than one during the dry season)
- Changes in the gravity field of Earth, reflecting changes in groundwater and surface water stores and soil moisture content (Chapter 5)
- Chlorophyll fluorescence of vegetation canopies.

Remote sensing (RS) has generated insights about processes, physical features, and ecological trends that were impossible to contemplate just 30 years ago. Some of the most dramatic insights include the following:

- Regional, continental, and global estimates of vegetation productivity
- Tracking deforestation of the Amazon
- Quantifying the areal extent of vegetation fires in remote regions
- Mapping and forecasting ocean primary productivity
- Mapping of land surface topography.

Much (but not all) of the insights that RS can provide in the fields of plant ecology, hydrology, resource management, conservation, agriculture, forestry, and land-use change arise because of the light (solar radiation) reflected from vegetation, land and ocean surfaces, and other features. In particular, the spectral characteristics of chlorophyll have played a prominent role in RS of terrestrial and oceanic productivity.

GREENING-UP OF THE TERRESTRIAL GLOBAL LANDSCAPE

The total area of leaf in a field, valley, region, country, continent, and globally, is a primary determinant of the productivity of that field, valley, region, country, continent, and globally, where productivity is defined as the rate of biomass produced per unit land area per unit sunlight available per unit time (there are other definitions, but this one works here). Local limits to productivity include low temperature (Chapter 4), insufficient water supply (aridity), and poor soil nutrient content. Leaf area index (LAI) is the key measure of the leaf area of a plot of land, at all scales. Leaf area index is the ratio of the number of square metres of leaf that sit above a square metre of land. An amazon rainforest might have an LAI of 6, which means that for every 1 m^2 of ground, there are, on average, 6 m^2 of leaf sitting above it within the under-storey, plus mid-storey, plus upper-storey, of the vegetation canopy. An arid site might have an LAI of 0.2, meaning that for every metre of ground, only 0.2 m^2 of leaf sits above it, on average.

Leaf area index is a key determinant of productivity simply because it determines how much light (how many photons) are absorbed by a plant. The efficiency of conversion of energy contained in photons to the energy contained in sugars produced by photosynthesis is very low, between 5 % and 11 %, making photosynthesis highly inefficient, so plants tend to increase their LAI if other factors are not limiting (for example, water supply, or N supply, or low temperature, are not constraining photosynthesis). Amazonian forests have a very high LAI because it is warm, there is plenty of sunlight available, water supply is abundant, and nutrients can be easily captured , retained and recycled efficiently.

Leaf Area Index can be calculated using remote sensing technology

Healthy green leaves absorb a large fraction of the incoming shortwave solar radiation, especially in the blue (about 400 – 500 nm) and red (about 600 – 690 nm) parts of the spectrum. They reflect most of the green wavelengths, hence their green colour. At wavelengths exceeding 700 nm, leaves are essentially transparent and absorb very little of the solar radiation above 700 nm (their reflectance and transmission accounts for almost all wavelengths exceeding 700 nm arriving at the leaf surface; Fig. 11). However, the percentage reflectance, which is less than 10 % for wavelengths 400 – 700 nm (the visible part of the spectrum) jumps very sharply, to about 50 % at 720 nm, for example. Leaves reflect a large fraction of near-infra-red wavelengths because the 3-D structure of the intersections of cell walls of adjacent cells of the leaf make them very good reflectors. Evolution has resulted in leaves reflecting a large fraction of the near-infrared wavelengths because (a) these wavelengths don't drive photosynthetic carbon uptake; and (b) absorbing these wavelengths would result in leaves becoming far too warm for their biochemistry to cope.

Herein lies the method used by satellites of measure the "greenness" of a land surface. Getting a handle on the vegetation greenness of the Earth's land surface turns out to be extremely valuable.

Many satellites contain spectroradiometers looking down at the Earth's surface. Satellite spectroradiometers can measure the wavelength of light reflected from specific areas of the Earth's surface. One of a widely used pair of spectroradiometers is the Moderate Resolution Imaging Spectroradiometer (MODIS) located on board the Terra satellite; the other one of the pair is on the Aqua satellite. Because MODIS sensors measure reflectance at both the near-infrared (NIR) and red wavebands, it is possible to calculate their ratio. With a slight tweak, it is then possible to calculate a very useful, widely used, and stable, vegetation index called the Normalised Difference Vegetation Index (NDVI), calculated as:

NDVI = (Reflectance of NIR – reflectance of red wavelengths)
÷ (Reflectance of NIR + reflectance of red)

NDVI can range between -1 and +1. Vegetation generally has an NDVI of between 0.3 and 0.8, with high reflectance within the NIR and low reflectance within the red wavebands. Soils tend to have an NDVI of about 0.1 – 0.2, snow and ice have an NDVI of about -0.1 to 0.1. The key point is the NDVI is very good at discriminating between free water (lakes, oceans), snow/ice, soil, and vegetation. A recent(ish) modification of the NDVI is the Reduced Simple Ratio (RSR), which uses the maximum and minimum values of the middle infrared waveband (3 – 50 μm) and shortwave infrared wavelengths (0.7 – 2.5 μm, or 0.9 – 1.7 μm depending on whom you believe). The RSR is very good at measuring the LAI of a vegetated surface. Entire books have been written about the derivation of many different vegetation indices and comparisons among vegetation indices. We don't want to bother about all that. What we want to know is: have there been any trends in terrestrial greenness over the past few decades, what do these trends tell us, and what might be causing them?

GREENNESS TRENDS OVER THE PAST 4 DECADES

Vegetation is the source of all the food we eat, the fibre we use for paper and clothing, and the oxygen (O_2) we breath (I am including ocean dwelling algae here; algae produce about 70 % of the O_2 in the atmosphere). Changes in LAI (greenness) influence the rate of water use of vegetation and this affects the amount of water that either runs-off into streams and rivers or goes to recharge groundwater stores (Chapter 5). In the absence of increased rainfall, when vegetation water-use persistently increases, the volume of water that can flow to streams and rivers or groundwater recharge declines, with knock-on effects for irrigation and river ecology. Clearly, understanding how global-scale greenness is changing is important to us.

In 1997 the first study to identify increased greening (of the northern hemi-sphere) was published, using NDVI data. Since then, numerous studies at regional-, continental- and global-scales have identified a "greening-up" of vegetation. For example, in 2009 a study of Australia, the driest of all permanently inhabited con-tinents, where vegetation is predominantly water-limited (which means that the availability of water limits the productivity of most regions of Australia), demon-strated that total vegetation cover increased by about 8 % between 1981 and 2006 and this was almost entirely due to an increase in the evergreen component of vegetation. Increases in LAI over the past 20–40 years have also been recorded in continental USA, India, China, and even the Arctic. In 2016, Zhu and co-workers demonstrated persistent and widespread greening over 25 – 50 % of the world's terrestrial surface and only <4 % exhibited a browning trend (declining greenness). An example of the types of results obtained is given in Figure 12, where the trend in LAI (not the absolute value of LAI) is predominantly one of increasing values, for the period 2000 – 2018.

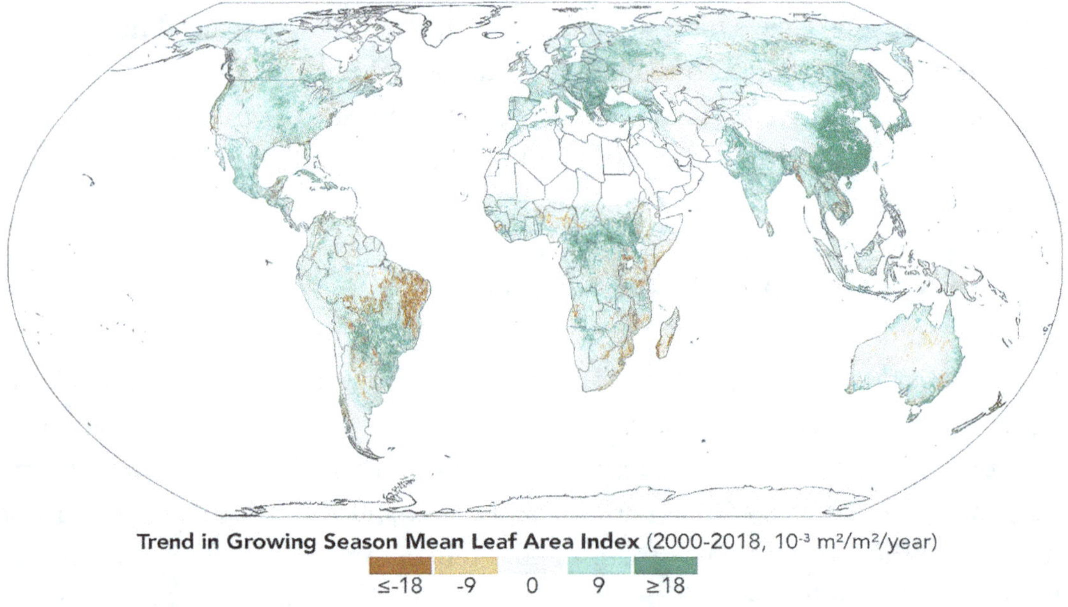

Figure 12: Global trends (not actual values) in growing season LAI 2000 – 2018. Much of the trend observed in China and India is the result of large-scale and long-term changes in land-use and the increasing use of irrigation and fertilisation for crops. From https://earthobservatory.nasa.gov/images/146296/global-green-up-slows-warming. NASA images are public domain.

WHAT HAS CAUSED INCREASED GREENNESS (INCREASED LAI) ON SUCH LARGE SPATIAL AND TEMPORAL SCALES?

Several variables interact to determine the LAI of a site. Soil nutrient content (especially N and P levels) and soil depth, temperature, and soil water availability are perhaps the most important natural (non-anthropogenic) variables, while atmospheric CO_2 concentration and land management practices (replacing forests with crops or *vice versa*, irrigation, and fertilisation) are the most important anthropogenic variables. The first two variables can be discounted because they haven't changed significantly in the past 4 or 5 decades (excluding the application of fertiliser in managed landscapes). What of the remaining variables?

Zhu et al., (Nature Climate Change 2016) combined satellite estimates of LAI and 10 global ecosystem models for the period 1982 – 2009 and established from the models that the CO_2 fertilisation effect accounted for 70 % of the observed trend in greening. Global ecosystem models are models of vegetation function and behaviour. In particular, they modelled the amount of carbon absorbed and released by vegetation and soil, and how these vary according to a number of variables, including temperature, the amount of light available, the amount of water available, and the concentration of atmospheric CO_2. The CO_2-fertilisation effect is the hypothesis that increasing concentrations of atmospheric CO_2 result in increased plant biomass because of the increased rate of photosynthesis of the canopy. This in turn is caused by the increase in supply of CO_2 to the primary CO_2-fixing enzyme within leaf cells. This definition of the CO_2-fertilization hypothesis is explicit about the feedback from increased storage of carbon in vegetation and the rate of accumulation of atmospheric CO_2. (I briefly discuss this feedback mechanism, or feedback loop, in the next paragraph.) Similarly, a large review published in 2020 (Walker et al., New Phytologist) suggested that the CO_2-fertilisation effect may account for about 50 % of the increase in greenness on the strength of the terrestrial carbon sink (a terrestrial carbon sink is an area where uptake of CO_2 exceeds release of CO_2 on an annual (or longer) time-frame). Oceans are also carbon sinks through the activity of corals, sea grasses, and micro- and macro-algae. In one of my own modelling studies using a highly mechanistic and highly detailed soil-plant-atmosphere model, I established that a 30 % increase in LAI (that is, significant landscape greening) would be expected in response to atmospheric

CO_2 concentration increasing from the pre-industrial level (about 275 ppm) to an enriched atmosphere (560 ppm).

There are two interesting feedback effects arising from increased LAI. The first concerns change in atmospheric CO_2 concentrations. Since the industrial revolution, human activities have increased atmospheric CO_2 concentrations by 48 % (an increase from 277 to 411 ppm, from 1760 to 2019), and this equates to an accumulation of 277 Pg C (277,000,000,000 tonnes of C) over that period. However, global-scale carbon accounting for that period estimates that total anthropogenic emissions to the atmosphere is about 645 Pg C. Therefore, 368 Pg of carbon are "missing" (645 − 277 = 368) and it can be safely assumed that this "missing carbon" is contained in the biomass of forests, woodlands, coral reefs, seagrasses, and oceanic micro- and macroalgae. Thus, the accumulation of more and more carbon in biomass (greening of the landscape) is reducing the rate at which the concentration of atmospheric CO_2 is increasing. Hence, global greening constrains the rate of increase of atmospheric CO_2 concentrations in a feedback mechanism. There is a second feedback loop also arising from greening of terrestrial landscapes. As LAI increases, the total volume of water transpired by plants increases and this cools the land surface, in the same way when we sweat, we cool our skin surface. Global greening since the early 1980s may have reduced global warming by as much as 0.2° to 0.25° C. Thus, the world would be even warmer if global greening was not occurring. Increased atmospheric CO_2 concentrations are stimulating plant growth, crop yield, and wood production, and enhancing the cooling of terrestrial surfaces. Gaia hypothesis anyone (see text box)?

Are there any field-based experimental observations of tree responses to increased atmospheric CO_2 concentrations or landscape greening?

Figure 13: An example of a FACE experiment, located in NSW, Australia (EucFACE). Six rings were constructed (4 are shown here) with 3 having CO_2 enriched concentrations within them. The central crane in each ring allows access to the canopy without interfering with the undergrowth and soil structure.

The Gaia Hypothesis

Homeostasis is a well-known biological property of living cells/organs/tissues/ individuals/colonies. All living systems have an optimum *range* of body temperatures, internal pH, salt concentrations, and other biophysical properties. Complex mechanisms have evolved to regulate, as much as possible, the steady-state condition of these variables, through space and time within cells/tissues/individuals/colonies.

The Gaia hypothesis can be viewed as an extension of this highly unifying biological principle to include the interaction of living systems with the *inorganic* (non-living) components of Earth in such a way as to produce the attributes of a self-regulating (that is, homeostatic) system that supports the existence of life on Earth. Changes in the abiotic (non-living) components of the global Earth system (for example, the concentration of CO_2 in the atmosphere; the temperature of the land-surface; salinity of the oceans) influence the behaviour of the biotic (living) components of the globe in a manner that can be *interpreted* as acting to return the global abiotic system to a state that is conducive to life on Earth. And, very importantly, *vice versa*; changes in the biota cause changes in the abiotic environment in ways that are supportive of the sustenance of life. The chemist James Lovelock was perhaps the first to scientifically formalise the Gaia hypothesis, named after the Greek goddess of Earth, mother of all life. The equivalent in Roman mythology would be Terra Mater (Mother Earth). In India, there is the Hindu Prithvi (or Privthi Mata, the Sanskrit name for Earth). The Gaia hypothesis proposes that Earth is a self-regulating complex system incorporating the biosphere, the pedosphere (soils and geochemistry), the atmosphere, and the hydrosphere. Dr. Lynn Margulis, a microbiologist, became closely associated with the Gaia hypothesis, applying her microbiological knowledge to expand the remit and potential mechanisms underpinning the Gaia hypothesis. Two examples of Gaia-like behaviour include the relative stability in Earth's surface temperature despite large changes in the amount of solar energy received by Earth over the past several billion years and the relative stability of the Earth's atmospheric composition, a feature that can be explained by the activity of different components of Earth's biota. Global-scale increases in fire frequency when atmospheric O_2 levels rise too high, has been said to be an example of a biotic/abiotic interaction acting to regulate Earth for habitation.

Entire conferences and books have been devoted to exploring, supporting, or attacking the Gaia hypothesis. It is probably true that at its heart, it is an unscientific hypothesis, not being open to falsification (but see the last chapter in this book on falsification) and it lacks a mechanistic basis linking the evolution of life and changes in biogeochemical processes. However, the "soft" version of Gaia, namely, that the evolution of life and biogeochemical processes are interactive, is, in my view, undoubtedly true. Whether we need the term Gaia hypothesis, however, remains a moot point.

It could be said that we are undertaking a global-scale experiment in testing the hypothesis that increased atmospheric CO_2 concentrations influence plant growth and C uptake. Technically, this is an unreplicated experiment (there is only one Earth) and it doesn't have a control treatment (a world without a change in atmospheric CO_2 concentrations). Therefore, it is not an acceptable scientific test of the hypothesis (Chapter 12). Consequently, we need a different approach. One such approach has been the development and construction of numerous FACE experimental sites. FACE stands for Free Air Carbon Enrichment and there are about 30 – 40 FACE experiments globally. An additional approach has been the long-term observational network of eddy covariance sites. I shall discuss both now....

There are 30+ FACE experiments currently in use globally. Figure 13 shows the Euc (short for Eucalypt) FACE site (for full disclosure, I was part of the team that wrote the report to the Australian government about the feasibility of establishing a forest FACE site in Australia). FACE sites consist, in essence, of a ring of pipes located at several heights above the ground, surrounding a patch of vegetation. This could be a small diameter ring around a patch of wheat, or some other crop, in which case it might only be 10 m in diameter and 2 m tall. Alternatively, it might encircle a patch of eucalypt woodland, or pine trees (Duke FACE, located at Duke University, USA) and FACE rings can be 20 – 30 m in diameter and 10 – 20 m high. CO_2 is released from the pipes in the upwind direction at a rate that maintains(ish) an enriched atmospheric concentration of CO_2 within the ring. A CO_2 monitor measures the CO_2 concentration at several heights at the centre of the ring and wind direction is also measured around the ring. Different FACE experiments have different set-points for the concentration of CO_2 required (for example, 550 ppm or 650 ppm). These experiments can run for 5, 10, 20 years and a large array of measurements can be made of soil processes (N mineralisation, for example) and plant processes (rates of stem growth, crop yield, rates of photosynthesis and respiration, for example). A set of control rings (no enrichment of CO_2 concentrations) and replication of rings are central features of the design of FACE experiments.

Some key findings of crop and forest FACE experiments are:

- Elevation of the concentration of CO_2 by about 200 ppm caused an approximately 18 % increase in crop yield under non-stress conditions. Leguminous and root crops showed a larger increase and cereals a smaller increase. Nitrogen deficiency and atmospheric warming (by about 2 °C) reduced the average increase to 10 %.
- Leaf-scale rates of light-saturated photosynthesis were increased, and the increase was broadly in agreement with theoretical predictions of leaf-scale photosynthetic responses.

- The response of biomass production is variable: in woodlands and forests that is an active biomass-accumulating phase, increased biomass production is generally observed, but more mature forests tend have a muted or zero response to CO_2 enrichment. Measurements of tree-ring width (a good proxy for biomass production) also do not provide a definitive response. However, although not universally observed, increased LAI (greening-up) of vegetation within FACE rings is frequently observed, in agreement with satellite data.

FACE experiments are complex, expensive, and generate an atmospheric CO_2 concentration that fluctuates in space and time within the ring's diameter. In contrast, the increase in atmospheric CO_2 concentration observed since the start of the Industrial Revolution has been gradual, incremental, and has not shown rapid fluctuations in time and space. The small number of FACE sites examining woodland and forest responses globally, makes them a small sub-sample of the Earth's ecosystems.

An alternative methodology uses sensors to measure the atmospheric CO_2 concentration and wind speed and wind direction 10 (or 20) times per second (10 or 20 Hz) and by using some pretty complicated maths, calculates the amount of CO_2 moving towards the canopy and the amount of CO_2 moving away from the canopy and takes the average of these values every 15 minutes. The sum of all these averages over 24 hours enables calculation of the influx (uptake) and efflux (loss) of carbon from the ecosystem, and by difference, establish whether a site is a net sink (uptake exceeds loss) or a net source (loss exceeds uptake) of carbon on a daily, seasonal, annual, and inter-annual timeframe. One of the real benefits of this approach is it is much cheaper than FACE experiments. Consequently, there are more than 500 sites with eddy covariance (EC) towers making these measurements (they are called eddy covariance towers because they measure the eddies in the wind above the canopy (using the 3-D sonic anemometer) and work out the covariances in wind speed, wind direction, and CO_2 concentrations). Figure 14 shows one of my two EC towers located in central Australia. When such towers have been functioning continually at a site for a decade or more, we can discern distinct trends in CO_2 fluxes, but also, we can analyse the fluxes in relation to important variables, especially rainfall, humidity, temperature, soil water content and vegetation structure. What have such analyses revealed over the past 10–20 years?

Figure 14: One of my eddy covariance towers in central Australia in a Mulga woodland (left image). The expensive gear that does all the measuring is at the top of the tower. The sensors include a 3-D sonic anemometer to measure wind speed and direction, and an open-path infra-red gas analyser to measure the concentration of CO_2 and H_2O (left-hand instrument).

Dr. Akihiko Ito, a researcher in Japan, took long-term (> 5 y) data from 118 EC sites around the world and obtained 1198 site-years of data (number of sites x numbers of years for which data were available at each site). As such, this represents a large, robust, and widely dispersed (with the exceptions of Russia and China) data set. Dr Ito demonstrated that for almost 60 % of sites, between 1992 and 2014, the terrestrial land surface (that is, soil plus vegetation) has been taking up more and more CO_2, that is, his analyses support the remotely sensed identification of "greening-up" discussed above. A key feature of EC data is that it integrates the behaviour of soils and all the vegetation within the "footprint" of the tower. The footprint of the tower is the area of land that can influence the signal measured by the gas analyser. It does not require the measurement of individual plants, species, or components of the ecosystem as it provides a truly integrated measure of C fluxes of a large patch of landscape.

Here is some weird stuff....

Over the past several decades or so, sea levels have been increasing by approximately 3 mm y^{-1} (Fig. 15). This is mostly because of the thermal expansion of water (water expands by about 0.2 % for an increase in temperature from 20 °C to 30 °C). Some of this increase is also because of the melting of ice from glaciers and Antarctic ice (remember that the Arctic is a floating ice mass where melting has no significant impact on sea levels).

What is remarkable is that in 2010/2011, there was a massive decline in sea levels (Fig. 15).

But what has this to do with EC and the greening of global land surfaces?

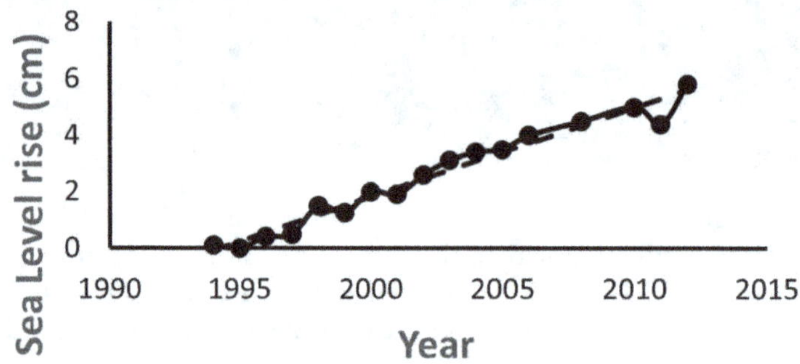

Figure 15: Sea levels have been steadily increasing for the past 30 year. However, in 2011 there was a large sudden decline in sea level. Redrawn from data in Boening et al. (2012, Geophysical Research Letters).

From the beginning of 2010 to mid-2011, global sea levels fell by about 5 mm. The volume of water lost from the oceans was deposited, predominantly, over Australia, South America, and Southeast Asia and this arose because of the change from El Niño in 2009/10 to La Niña in 2010/2011. La Niña is associated with a decrease in upper ocean temperature in the western tropical Pacific and higher temperatures in the western tropical Pacific, resulting in a large increase in evaporation of ocean water from the western tropical Pacific and consequential large increases in rainfall across eastern and central Australia. Data from satellite altimeters and the twin GRACE satellites (Chapter 5), when combined, were able to detect both the decline in sea levels and the increase in surface water content across a large fraction of Australia. During the La Niña period, Australia received about 55 % more rainfall than the long-term annual average of about 450 mm. Because much of Australia is water limited (vegetation productivity is limited by the availability of water), such a large increase in rainfall and hence soil water content should have resulted in a significant greening of the continent. And indeed, it did. Central Australia is particularly arid and my EC tower, situated above a Mulga low woodland not far from Alice Springs, showed that the Mulga had a net ecosystem productivity (NEP; input of C via photosynthesis minus loss of C via respiration) of pretty close to zero in 2010 but NEP increased to 259 g of C fixed per square metre of land surface in 2011. This was in response to the massive increase in soil water content, as well as the increase in atmospheric humidity and slightly cooler temperatures, both of which were of benefit to the Mulga trees. Globally, the size of the terrestrial sink for C in 2011 was estimated to be 54 % *larger* than the average for the preceding decade, equivalent to 40 % of global emissions of C from burning fossil fuel. Who would ever have predicted that changes in sea levels, patterns of rainfall in the tropical Pacific, and greening of Australia's arid ecosystems could all be linked so directly?

To summarise this chapter, the following can be said:

The temperature of an object determines the dominant wavelengths of light emitted by that object and because the sun is hot, it emits a lot of photons between 250 and 2250 nm, with a peak between about 400 and 700 nm, which corresponds to the wavelengths that humans can see, and which plants use to drive photosynthesis, and therefore carbon gain.

The sky is blue for most of the sunlit period because photons are scattered as they bump into gas molecules and dust and aerosols in the atmosphere and more blue photons are scattered downwards towards Earth than red photons. Just after sunrise and before sunset, the sky can look red because of the increase in distance that photons must travel through the atmosphere, especially the lower atmosphere, where most of the dust and atmospheric gases are located and blue photons are scattered away from our direct line of vision, hence the photons we see are enriched in red and yellow photons.

Grass is green because chlorophylls in leaves absorb red and blue photons well but absorb very few green photons. However, chlorophyll can look bright red because when excited electrons fall back to the unexcited orbital, they release red photons. Interestingly, the structure of chlorophyll and haemoglobin, foetal haemoglobin, and leghaemoglobin (found in roots of leguminous plants) are very, very, similar.

The greenness of the Earth's terrestrial surface can be quantified using remote sensing, using sensors mounted on satellites, and such measurements have detected the global phenomenon of "greening-up", with increased greening apparent across large areas of Earth, including across Australia, the USA, China, India, and the Arctic. While some of this (especially in China and India) is the result of changes in crop cover and the increase in irrigation and fertiliser use in these regions, much of this greening-up has been attributed to the increase in atmospheric CO_2 levels over the past 30 years and therefore a natural, overall increase in leaf area index in many parts of the world. Field measurements using FACE experiments and eddy covariance towers have generally supported the greening-up detected in space.

Finally, we saw a direct link between large regional-scale weather patterns (El Niño and La Niña), sea level changes and vegetation greening, with Australia as a key case study, highlighting the important role of Australian arid-zone vegetation to interannual variability in global atmospheric CO_2 levels.

THE END

CHAPTER 11

The surface temperature of Earth is currently increasing (global warming is a reality). This is not a contentious statement. Similarly, global atmospheric CO_2 concentrations are increasing, that is also clear. However, which came/comes first – changes in mean surface temperature or changes in atmospheric CO_2 concentrations? If increases in temperature precede increases in atmospheric CO_2 concentration, is there a hypothesis that might explain how temperature increased prior to the Industrial Revolution? And, what has caused increases in atmospheric CO_2 observed that occurred after the start of the Industrial Revolution (about 250 years ago)? In answering these questions, I challenge the prevailing paradigm (with evidence), and [daringly] propose that our current response to global warming and climate change is not evidence-based, may be misguided, and the lack of robust cost-benefit analyses is leading us down the wrong response pathway.

Too confronting?

There appears to be much confusion in the public's mind, and in media reports, concerning the difference between weather and climate. Such confusion feeds directly into misunderstandings and misinterpretations of recent weather events and results in untrue statements about the evidence for climate change.

Let me state at the outset that I have received extensive research funding in the past to study plant responses to increased atmospheric concentrations of CO_2. I cannot, in any sense, be labelled a "Climate Change Denier". Some indisputable facts:

- Climates have been changing for thousands of millions of years.
- Global average temperatures have been increasing for the past 200 years (approximately).
- Increased atmospheric levels of CO_2 do indeed increase surface temperature.

The purpose of this chapter is not to disparage the Science of climate change (the subject of Chapter 12), but to address those three issues raised in the first paragraph. I also briefly consider whether the strategy of decarbonising the West is in our best interests in today's complex geopolitical climate, or is there a better approach to tackling the problem of climate change?

WHAT IS THE DIFFERENCE BETWEEN WEATHER AND CLIMATE?

Weather forecasts in newspapers often refer to temperature, rainfall, wind speed, humidity, and solar radiation levels/UV levels for today, tomorrow, and sometimes for the upcoming 7-day period. Some weather forecasts stretch themselves out to a 1-month timespan. Here lies the answer to this first question. Weather is what we experienced yesterday, are experiencing today, and might experience tomorrow, or next week. The timescale is measured in days, perhaps weeks.

Climate, in contrast, is the *average* temperature, rainfall, humidity, and wind speed (for example), for periods *exceeding* 30 years. A 30-year average of climate variables can be calculated for a particular location (Washington, D.C., for example) or across a larger spatial scale (California or Australia, for example). Importantly, we must differentiate long-term averages of past observational (measured) data, and estimates of averages for future climate scenarios, which rely on increasingly complex global circulation models. For perspective, if printed on paper, the computer code for a typical global circulation model would cover about 20,000 pages of text.

HOW ARE WEATHER FORECASTS PRODUCED?

Weather forecasts are produced using numerical weather models (NWMs). Numerical weather models are mathematical descriptions of two fluids: the atmosphere and the oceans. Because oceans supply water vapour to the atmosphere, and supply/absorb heat and CO_2 to/from the atmosphere, NWMs must account for (that is, calculate) their behaviour, especially the movement of air and water masses, the temperature and density of these masses, and the movement of heat within and between both masses, and all the interactions among these variables. Some basic and well-understood equations that describe the conservation of mass, momentum, and energy, and the ideal gas law (which describes the relationship between the pressure, volume, and temperature of a gas) form the overarching equations used in NWMs. However, of themselves, these equations are insufficient to adequately predict weather. There are more than 20 physical processes that happen at small(ish)-scales (*ca* 10 m to 50000 m) that can't be represented in the overarching equations because they are insufficiently understood, occur too rapidly, or require too much computational power, to include at the spatial/temporal scale at which they occur (small-scale and short-term). These poorly understood processes include, for example, the emission of longwave radiation (especially infra-red radiation) from the Earth's surface, cloud microphysics, turbulence in the atmosphere, incoming (shortwave) solar radiation, and transpiration (water loss) by vegetation. To overcome this problem, NWMs include a parameterization scheme, where these processes are summarized using reasonable physical *approximations* and *statistical representations*.

An important component required to run a NWM is current weather at different heights above the ground, with as high a spatial resolution as can be achieved. Current weather information is provided by radiosondes (sensors to measure and transmit important weather variables). These radiosondes are often sent up into the atmosphere beneath weather balloons, or are located on satellites or in ground-based weather stations.

Numerical weather models represent the Earth's surface (land and sea) as a 2-D grid of squares that range in size from 5 km square to 300 km square. The third dimension, vertical height above the Earth's surface, is also included, creating a grid box, that approximates a cube, but with curved sides.

In the 1950's it took 5 days to do the computations, by hand, to generate a weather forecast 6 h into the future. Today, supercomputers continually process incoming meteorological variables within NWMs and produce forecasts that are increasingly accurate, to about 6 days ahead. It is unlikely that accuracy beyond about 10 days can be much improved, even with bigger/faster computers, because of the chaotic nature of weather and because of the low spatial and temporal resolution of incoming data. Once NWMs are applied to the globe they become Global Circulation Models (GCMs), which can be used to model patterns in global circulation of the atmosphere and oceans, and derive weather patterns and future scenarios of climate change. One of the major features of GCMs as applied to future climates is the incorporation of the effects of greenhouse gases on the energy balance (the difference between incoming and outgoing energy) of Earth. Changes in energy balances impact directly on weather patterns and climate.

GREENHOUSE GASES, RADIATIVELY ABSORBENT GASES, AND TEMPERATURE

Greenhouse gases, or as I prefer, radiatively absorbent gases (RAGs), include the gases CO_2, CH_4 (methane), water vapour, and N_2O (nitrous oxide), among others. The key feature of these gaseous molecules is that they are diatomic. They contain 2 different atoms (C and O, C and H, H and O, N and O) and this property means that they are very good at absorbing infrared (IR) radiation (light in the waveband 700 nm to 1 mm). They absorb IR energy (which we experience as heat) when the bonds linking the two different atoms bend or stretch, and the molecules vibrate (because of the extra energy they now contain). A short time later, the molecules release a photon of IR radiation. A proportion of these photons are directed downwards and can warm the surface of Earth. A proportion are absorbed by other RAG molecules, and a proportion are directed out to space. Similarly, a proportion of the IR radiation emitted by all objects on the surface of Earth (soil, rocks, people, oceans, plants, animals) is absorbed by RAGs and a proportion is emitted out to space. Herein lies the "greenhouse effect". As the concentration of RAGs in the atmosphere increases, the number of IR photons absorbed by the atmosphere and the number of photons reemitted to Earth from those RAGs, increase, leading to a warming of the atmosphere and the Earth's surface. The atmosphere is acting as a global blanket, insulating the planet. This is vitally important. In the absence of RAGs in the atmosphere, the *average* temperature of the Earth's surface would fall by about 30 – 35 °C (from about +15 °C to about -20 °C). Most of Earth's water would freeze (see Chapter 3 for a discussion on this point).

The Swedish scientist Svante August Arrhenius published a paper in the Philosophical Magazine and Journal of Science in 1896 quantifying the contribution of CO_2 to the greenhouse effect and speculated on the link between CO_2 concentrations and climate. It took several decades for a mechanism to be established linking CO_2 concentrations and IR absorbance.

Different RAGs have different global warming potentials (GWP). The GWP is a measure of the heat absorbed by a greenhouse gas, expressed as a multiple of the heat that would be absorbed by the same mass of carbon dioxide. The GWP of CO_2 is, by definition, 1 because it is the reference gas. Methane has a GWP of about 30 times that of CO_2 over 100 years, nitrous oxide has a GWP of about 280 times that of CO_2 for a 100-year timescale. Thus, despite the concentration of methane being about 1890 ppb (1.89 ppm), that is, about 0.5 % of that of CO_2 (currently > 424 ppm), it remains an important RAG because its GWP is 30 times larger than that of CO_2.

WHAT HAS CAUSED INCREASED ATMOSPHERIC CONCENTRATIONS OF CO_2?

Fire has probably been used by humans for between 400,000 and 800,000 years, initially using wood and dried dung as the primary fuels. Humans have burnt coal for more than 3000 years, but the world's population was fewer than 100 million people 3000 years ago, and not everyone was burning coal. It wasn't until the Industrial Revolution (generally agreed to have started in the 1760s in the UK, although examples of the design and manufacture of metal machines predate this) that coal and oil became increasingly used. The first Newcomen atmospheric engine (an engine where the power stroke of the piston is driven by atmospheric pressure) was installed near Dudley Castle in the UK in 1712, to pump water out of coal mines. James Watt's steam engine was finalised in the mid-1760s and is often considered to be the hall-mark of the start of the Industrial Revolution.

Several key features characterised the Industrial Revolution (IR). The first was a rapid increase in the use of machines to replace muscles in manufacturing, agronomy, and transport. The second was a dramatic increase in the use of fossil fuels, starting with coal, for domestic and industrial use (to make iron, steel, and steam). Third, population numbers increased dramatically, especially in the UK and Europe, and then elsewhere. The population of England and Wales, for example, increased from about 6 million in 1750 to about 18 million in 1851.

Atmospheric CO_2 concentrations were relatively stable at around 280 ppm from the year 200 AD to about 1880 AD and only increased to about 300 ppm by the early 20th century (currently, global average atmospheric CO_2 concentrations exceed 424 ppm).

How do we know the CO_2 concentration in the atmosphere increased in the 18th and 19th centuries? There were no electronic gas analysers available at the time.

CO_2 concentration in bubbles of air trapped in ice in Antarctica (and other places) can be measured once the ice is melted in sealed vacuum containers. One such set of ice cores has been collected at Law Dome, East Antarctica, between 1987 and 1993. Ice cores are collected, kept frozen, and then melted within a vacuum chamber. The gas released was collected and CO_2 concentrations measured using modern analytical techniques. It is also necessary to obtain accurate dating of the ice from different depths. The deeper the sample, the older the bubble. Ice from different depths was melted and the cores were dated by counting the annual layers in oxygen isotope ratio (the ratio of ^{16}O to ^{18}O in H_2O; ^{16}O is the most common isotope of oxygen, ^{18}O is a rare isotope of oxygen. The ratio of the two is correlated with surface temperatures and is therefore commonly used as a proxy of surface temperature). From these, and other measurements, seasonal cycles were clearly visible and by counting the number of cycles as a function of depth, a dating accuracy of ±2 years at 1805 AD (for example) was obtained, which was pretty impressive.

HOW HAVE ATMOSPHERIC CO_2 CONCENTRATIONS AND GLOBAL SURFACE TEMPERATURES CHANGED DURING THE EARLY-TO-LATE 19TH CENTURY?

From the start of the Industrial Revolution (1760) through to about 1910, fossil fuel consumption and therefore annual rates of CO_2 emissions, were very low. Consequently the *rate of increase* in atmospheric CO_2 concentration was very low. In contrast, despite the low rate of increase in CO_2 concentration to 1910, surface temperatures increased rapidly throughout 1830 to 1910 (Fig. 1). Beyond 1910 (ish), atmospheric CO_2 concentrations rapidly increased in response to increased CO_2 emissions, and the rate of rise in surface temperatures also increased (Fig. 1).

Focusing on the key period prior to 1910 in Figure 1 leads one, inexorably, to the hypothesis that increased temperature preceded increased atmospheric CO_2

concentration. Indeed, have we currently got "cause-and-effect" the wrong way around?

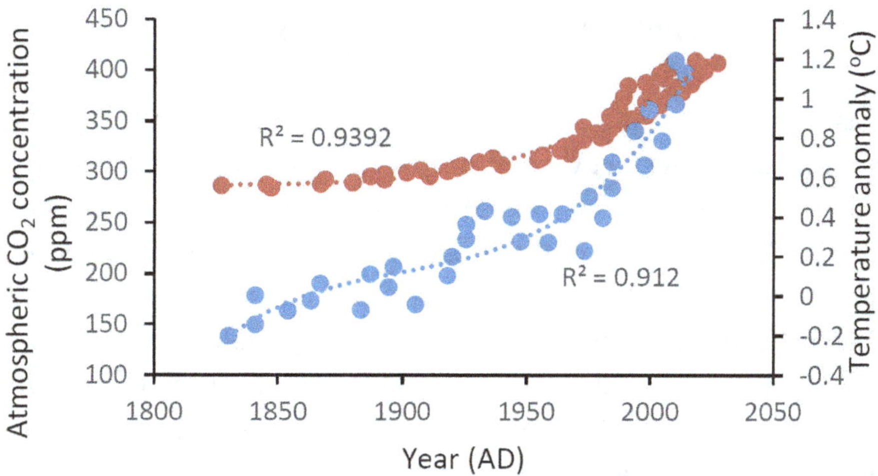

Figure 1: Changes in atmospheric concentration (red circles) were very small from 1760 to 1900 (the first 65 y are omitted for clarity). In contrast, surface temperature anomalies (the difference between global mean surface temperature in any given year and the average surface temperature of a given reference period; blue circles) were increasing from the early 19th century. Data from https://www.climate-lab-book.ac.uk/2020/2019-years/ and a CSIRO CO_2 data source. The temperature reference period is 1850 – 1900.

An aside on temperature anomalies

In studies of climate change, *temperature anomalies* (or *temperature changes*) are more important than absolute temperature. A temperature anomaly (or temperature change) is the difference between a reference temperature (usually the average across a 30-year window) and the average temperature for any given year (Fig. 1). A positive anomaly means that the temperature observed in any given year (or set of years) was warmer than the reference period. A negative anomaly indicates the observed temperature was cooler than the reference. When calculating an average of absolute temperatures, things like station location or elevation have an effect on the data. However, by using anomalies, such factors become less important.

Using anomalies also helps minimise problems when stations are added, removed, or missing from a monitoring network.

CAUSE AND EFFECT: WHICH COMES FIRST – INCREASED SURFACE TEMPERATURE OR INCREASED ATMOSPHERIC CO_2 CONCENTRATION?

A CAUTIONARY TALE FROM PALAEOCLIMATIC STUDIES

The journal *Nature* is one of the top four journals in the world. In 1999 Petit and colleagues examined lags between atmospheric CO_2 concentrations and surface temperatures over the past 420,000 years by sampling ice cores drilled at Vostoc, in Antarctica. The data showed that declines in atmospheric CO_2 concentrations lagged declines in surface temperatures by several thousand years. Periodicities of about 20,000, 40,000 and 100,000 years in climate variables were readily apparent in the data extracted from the ice cores. These periodicities are the hallmark of the well-known Milankovitch cycles (Chapter 4). Milankovitch cycles describe the effects of changes in the Earth's movements on Earth's climate over thousands of years. The Serbian geophysicist and astronomer Milutin Milanković hypothesised that variations in eccentricity, axial tilt, and precession (defined a few sentences down), combined to result in cyclical variations in inter-annual and latitudinal distribution of solar radiation arriving at the Earth's surface, and that these strongly influence Earth's climate. Eccentricity is a measure of the extent to which the Earth's orbit in any given year is circular or elliptical, which determines the annual average distance of Earth from the Sun, and therefore the amount of solar radiation received by Earth, thereby affecting surface temperatures. Major and minor components of cycles of eccentricity combine to give a periodicity of about 100,000 years. Earth rotates around a N-S axis that is tilted with respect to the orbital plane (the obliquity of the ecliptic), varying between 22.1° and 24.5° over a cycle of about 41,000 years. Currently the axis tilts about 23.44°, and this is slowly declining. Decreasing tilt decreases the amplitude of the seasonal cycle in sunlight received by Earth. Axial precession refers to the fact that the N-S axis "wobbles" so that the axis describes a circular orbit – eventually, for example, Polaris will not be the North Polar Star because the N-S axis won't point to it. Precession has a periodicity of about 25,700 years. **Petit and colleagues concluded that changes in atmospheric CO_2 concentrations amplify the *initial* effects of orbital forcings on surface temperatures, but changes in atmospheric CO_2 concentrations can follow, and not cause, changes in temperature (see below).**

The journal *Science* is also one of the global top four science journals. In 1999 Fischer and colleagues examined ice core records of atmospheric CO_2 concentration during the last 3 glacial terminations (all within the past 300,000 years) and concluded the following:

a. high-resolution records from Antarctic ice cores show that CO_2 concentrations increased by 80 to 100 parts per million by volume, 600 ± 400 years *after* the warming of the last three deglaciations;

b. the time lag of the rise in CO_2 concentrations with respect to temperature change was in the order of 400 to 1000 years during all three glacial-interglacial transitions;

c. during the penultimate warm period, CO_2 concentrations reached their maximum 400 ± 200 years later than Antarctic temperatures.

Unfortunately, these paleoclimatic studies do not explain *modern-era patterns in temperature and CO_2 concentration* because the timeline is too short for Milankovitch cycles to be driving temperature. *What they do demonstrate, however, is three key principles.* The first is that changes in CO_2 concentrations **can** precede changes in temperature. The second is that CO_2 concentration **can** amplify small changes in temperature into larger changes in temperature. The third is CO_2 concentrations and temperature *are linked in a feedback loop*. This feedback loop between temperature and atmospheric CO_2 concentrations means that changes in one cause changes in the other. More of this later.

Recent evidence

Is there any evidence that lags between temperature and CO_2 concentration are apparent in more recent times? Well, Figure 1 certainly shows a large increase in temperature prior to large increases in CO_2 concentration in the early-late 19th century. This is, I feel, particularly emphatic. What of the 20th/21st centuries?

A paper with the snappy title "Atmospheric temperature and CO_2: hen-or-egg causality?" (Kootsoyiannis and Kundzewicz, *Sci*, 2020) explicitly examined the possibility of increased surface temperatures causing increases in atmospheric CO_2 concentrations. These researchers used reliable instrumental temperature data and

CO_2 concentration for the period 1980 – 2019 and concluded the following key points (Fig. 2):

- While it is undoubtedly true that increasing atmospheric CO_2 concentrations can result in increased surface temperatures, the results demonstrate that the dominant direction of causality (during the period 1980 – 2019) was the reverse of the dominant paradigm, that is, *increasing temperatures resulted in increased atmospheric CO_2 concentrations, not the reverse.*
- Between 1980 – 2019, changes in atmospheric CO_2 concentrations followed changes in surface temperature. Figure 2 demonstrates this well (redrawn from Kootsoyiannis and Kundzewicz, 2020).

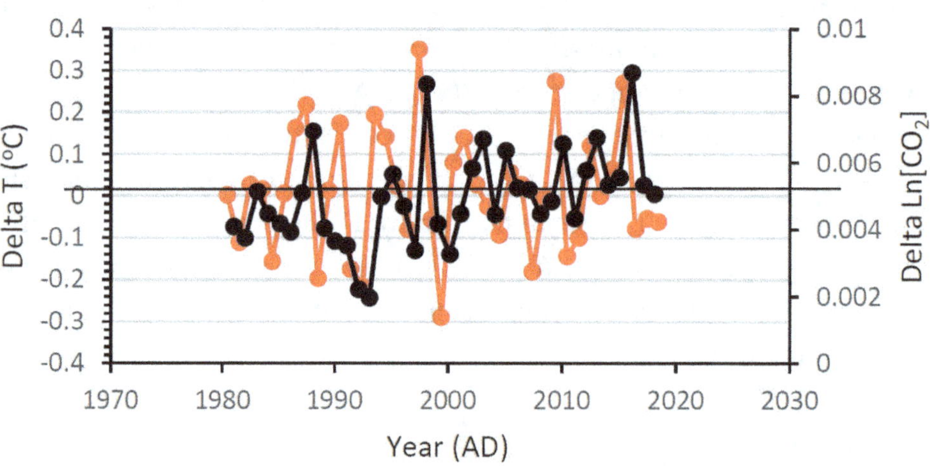

Figure 2: Changes in temperature (Delta T; red line) between 1980 and 2020 precede changes in atmospheric CO_2 concentrations (black line) by 6 to 18 months. Redrawn from Kootsoyiannis and Kundzewicz, Sci, 2020. Ln means natural log (or $\log_e$) rather than log to base 10.

Is there a plausible mechanism for the reverse paradigm?

A plausible mechanism for the hypothesis that increased surface temperature *causes* increased atmospheric CO_2 concentrations proposes that increased surface temperatures cause soil respiration to increase resulting in a large loss of CO_2 from soil to the atmosphere (Kootsoyiannis and Kundzewicz, 2020). Importantly, soil respiration produces about 100 Gt C y^{-1}, compared to annual global anthropogenic emissions of about 10 Gt C y^{-1}. Small increases in soil respiration are therefore able to significantly increase the rate of CO_2 accumulation in the atmosphere.

Having a plausible mechanistic explanation for reversal of the presumed chicken-and-egg relationship lends considerable support to subversion of this dominant (exclusive?) paradigm. The judicious next question is: is there any evidence to support the hypothesis that increased surface temperature results in increased soil respiration? **The answer is a resounding yes,** as demonstrated in multiple laboratory and field experimental studies. Three conclusions from three large meta-analyses of global data will suffice. The first is from Chen and colleagues (Chinese Science Bulletin, 2013), who concluded the following:

- "Annual global rates of soil respiration increased from 1970 to 2008 (the period of the study) and this increase was correlated directly with global temperature anomalies, suggesting that *interannual variation in temperature was responsible for inter annual variations in predicted soil respiration.*"

Similarly, Yan and colleagues published a review of global data on soil carbon efflux (J. Soils and Sediments, 2019) and concluded the following:

- "Across all terrestrial ecosystems, warming reduced soil organic carbon content by 4.96 %, stimulated soil microbial biomass C, soil respiration, and heterotrophic respiration by 6.30, 14.56, and 8.42 %, respectively."

These results indicate that warming induces accelerated transition of soils from C sink to C source. Furthermore, they show the potential for global warming effects to exacerbate the positive feedback loop in terrestrial ecosystems.

Finally, Lei and colleagues published (Nature Communications, 2021) a modelling study that used data from 693 studies and stated:

- "Our analysis directly addresses the long-held concern about the positive land C-climate feedback that could accelerate planetary warming in the 21st century, which is critical for ecological forecasting and climate policy-making. Given the huge impacts of warming on large soil C storage in cold regions, the stronger increase of R_S (soil respiration) in high latitudes warrants more efforts…"

The positive land C-climate feedback they are referring to is that warmer temperatures lead to more soil respiration, which leads to higher atmospheric CO_2 concentrations, which leads to increased temperatures, which leads to….etc.

Looks like the mechanistic interpretation of Koutsoyiannis and Kundzewicz might have legs.

If we accept, for now, that increased temperatures can precede increased atmospheric CO_2 concentrations, we must ask, what might have kick-started the rise in temperature apparent in Figure 1? The answer *could* be an unfortunate sequence of megadroughts across large parts of the world. A megadrought is a prolonged period of drought extending across very large areas. For example, in 1780 – 1810 there was a long sequence of droughts and famines across much of India, causing more than 11 million deaths, and possibly the deindustrialization of India. Other examples include the Strange Parallels Drought (1756 – 1768) which affected most of India, Thailand, Laos, Vietnam, Cambodia and western Russia, while the East India drought (1790, 1792 – 1796) affected north and eastern India, Burma, Malaysia, Japan, and much of China. Most of eastern China was affected by drought between 1770 and 1790. Europe did not escape untouched, with the megadrought of 1770 – 1840, with the driest period occurring 1770 – 1820. Coincident megadroughts were recorded in the south west of northern America and the southwest of southern America in the late 18th century.

The thing about megadroughts is that they cause surface temperatures to increase across vast areas of the globe, for many years. Interestingly, continental USA is in the grip of a pan-continental megadrought (2000 – present?). However, before we rush to ascribe climate change as the cause of this particular megadrought, we should note that pan-continental droughts have occurred across the USA in 12 % of all years since the 10th century, and multi-decadal pan-continental droughts occurred across the USA in the 12th and 13th centuries, well before the IR.

If you find it hard to believe that a sequence of unpredictable events such as a series of megadroughts could alter global temperatures, consider that fact that the Little Ice Age (*circa* 14th to early 19th century), which followed the Medieval Warm Period, *could* have been triggered and sustained by a series of volcanic eruptions in

1257, 1268, 1275, 1284, 1452/3, 1580, 1600, 1660, 1783 and 1815 (the eruption of Mt. Tambora, the largest eruption in recorded human history; Miller et al., 2012, Geophysical Research Letters). Vast quantities of ejecta, especially ash and SO_2 (which form particulates in the upper atmosphere) can reflect sufficient incoming solar radiation to cool Earth for many years.

WHY DO WE CARE ABOUT THE CAUSE-AND-EFFECT RELATIONSHIP?

Because if surface temperatures are the *initial* cause of increased atmospheric CO_2 concentrations, we are currently treating a symptom. Just as Panadol treats a symptom of a viral or bacterial infection, namely a fever, without addressing the cause of the fever, it makes us feel better, but doesn't treat the infection.

Let us propose the hypothesis that significantly cutting anthropogenic CO_2 emissions will significantly reduce the rate of increase in atmospheric CO_2 concentrations. This is, of course, the basis for the head-long rush to decarbonise the West.

It turns out that we ***have fairly recently tested this hypothesis experimentally, three times,*** in 1979 – 1983, 2008 – 2010, and 2019 – 2021.

Lets unpack that last sentence a little……..

The following two graphs (Figs. 3, 4) show annual global CO_2 emissions since 1970 (Fig. 3) and monthly changes in atmospheric CO_2 concentrations measured at the top of a mountain on Mauna Loa in Hawa'ii (Fig. 4). If you are wondering why atmospheric CO_2 concentrations show an annual oscillation (minima in the northern summer, maxima in the northern winter; Fig. 4), this is because in the northern summer, temperatures are high (relative to winter), sunlight is more abundant (relative to winter) and deciduous trees and summer crops have a large canopy of leaves absorbing vast quantities of CO_2 from the atmosphere. In winter, evergreen conifers take up very little CO_2 from the atmosphere (Chapter 4), deciduous trees don't have a canopy, daylengths are short and temperatures are low (relative to summer). In addition, in winter, more fuel is burned to keep everyone warm.

The reason for measuring atmospheric CO_2 concentrations on the top of a mountain in Hawa'ii, in the Pacific Ocean, is (a) to be as far away as reasonably possible from major industrial sources of CO_2, and (b) to be moderately high to capture well-mixed air.

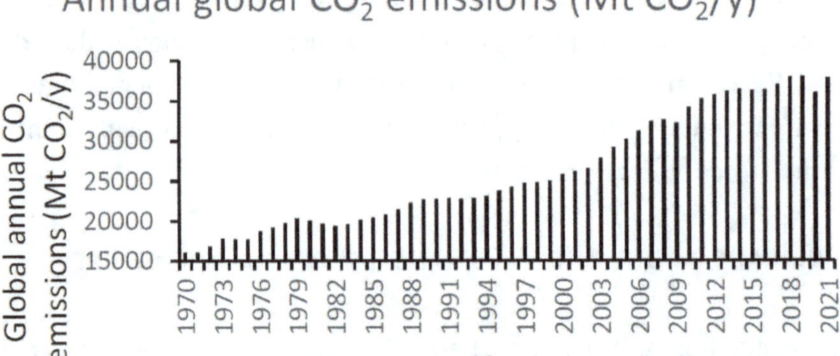

Figure 3: Annual global CO$_2$ emissions have been steadily rising since the Industrial Revolution. Here data from 1970 to 2021 are plotted. Redrawn from the EUROPEAN COMMISSION: EDGAR - Emissions Database for Global Atmospheric Research https://edgar.jrc.ec.europa.eu/report_2022

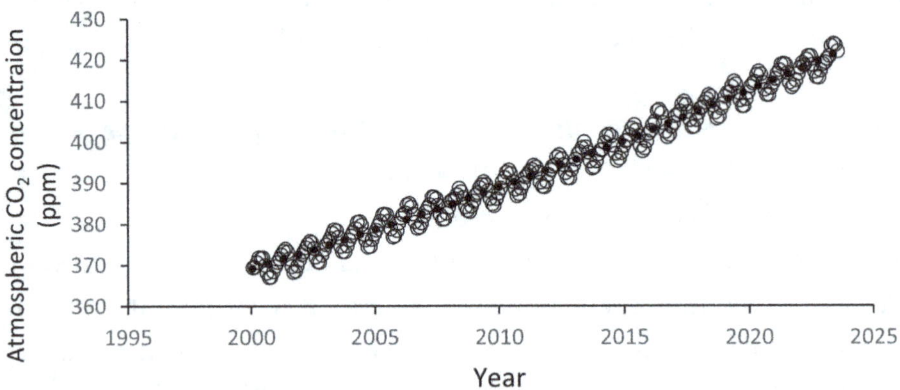

Figure 4: Atmospheric concentrations of CO$_2$ have been steadily rising since the Industrial Revolution. Here data from 2000 to 2023 are plotted (so that the seasonal cycle can be shown clearly). Redrawn using data from the Global Monitoring Laboratory: Earth System Research Laboratories. https://gml.noaa.gov/ccgg/trends/data.html

Several features apparent in Figures 3 and 5 (also in Fig. 4 but it is harder to see, which I why I plot subsets of Fig. 4 in Fig. 5, and use annual mean values rather than monthly values) provide a test of the hypothesis that changes in global CO$_2$ emissions (either reductions or a cessation of the preceding annual increases) result in significant declines in *the rate of increase* in atmospheric CO$_2$ concentration. I focus on three periods: 1) the early 1980s recession that affected much of the world between early 1980 and 1982. This recession may have been the most severe recession since World War II; until 2) the 2007 – 2008 financial crisis; and 3) the COVID-19 pandemic of 2020.

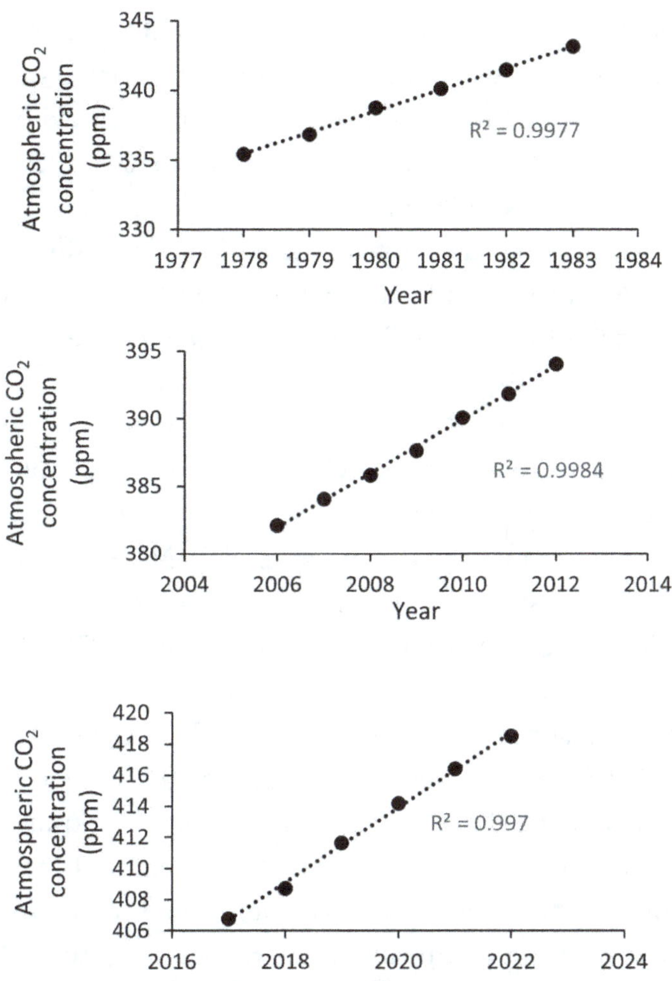

Figure 5: Atmospheric concentrations of CO_2 have been steadily rising since the Industrial Revolution. Here, data for three periods that include years before, during and after significant declines in global C emissions (from Fig. 4) are presented. Data are from the NOAA Global Monitoring Laboratory: Earth System Research Laboratories. https://gml. noaa.gov/ccgg/trends/data.html

1. During the early 1980s, the widespread and severe recession caused global C emissions to decline for a period of 3 y (Fig. 3). To test the hypothesis that reduced global emissions of C will reduce the rate of rise of atmo-

spheric CO_2 concentrations, I plotted the atmospheric concentration of CO_2 for the period 1976 – 1987 (upper panel, Fig. 5), thereby capturing the years before, during and after the recession. It is readily apparent that the slope of linear increase in atmospheric CO_2 concentrations did not change in response to the recession and the associated drop in global CO_2 emissions. Note that over 3 years, C emissions declined by 4.2 %, a significant decline in a short period. This is equivalent to the CO_2 emissions of Canada plus the UK in 2020.

2. In 2007 – 2010, an almost constant rate of CO_2 emissions was apparent for 2007 and 2008, followed by a small decline in 2009 (caused by the global financial crisis of 2008/9), followed by a large increase in emissions in 2010 (Fig. 3). Despite this pattern of change in annual emissions over a 4-year period, atmospheric CO_2 concentrations showed no change in slope for the period 2006 – 2012, that is, CO_2 concentrations continued to increase (middle panel, Fig. 5) at the same rate before, during and after the global financial crisis and its associated fall in global CO_2 emissions.

3. The final and largest test was the year 2020. A large decline (5.4 %) in annual CO_2 emissions occurred in 2020 because of COVID-19 restrictions on travel, reduced consumptive demand and reduced industrial activity. However, atmospheric CO_2 concentrations kept rising at the usual rate for 2017 – 2023 (a single slope fits the data for the years before, during, and after the pandemic; lower panel, Fig. 5). I must acknowledge that it was the paper by Kootsoyiannis and Kundzewicz (2020) that first alerted me to this important observation (that the decline in emissions in 2020 was very large but it had no impact on the rate of increase in atmospheric CO_2 concentration). The decline in global emissions in 2020 is equivalent to the 2021 annual emissions of Australia plus Canada plus Germany plus the UK plus Kuwait (or Australia plus Canada plus France plus Germany plus Sweden plus Switzerland). This is not an insignificant decline in emissions. Critically, we must ask: where did all the extra CO_2 come from during 2020 to cause the rate of increase in atmospheric concentration to remain unchanged, despite a 5.4 % decline (2020 compared to 2019) in anthropogenic emissions? I shall come back to this question towards the end of this chapter.

The evidence in Figures 3, 4, and 5 do not support the hypothesis that we are able to reduce the rate of increase in atmospheric CO_2 concentration through significant reductions (e.g., 5.4 %) in annual emissions. We have not, *ever*, achieved a deliberate, planned, reduction in global annual CO_2 emissions of the magnitude observed in 2020. And if we did reduce emissions by more than 5 %, the evidence in Figure 5 suggests it would have no impact on atmospheric CO_2 concentrations. Thus, massive reductions in fossil fuel consumption by Australia and a host of other economies will not have any impact on the rate of increase in atmospheric CO_2 concentrations.

One last statistic. Between 2010 and 2021, global annual emissions of CO_2 increased by about 11 %. There was no significant change in the annual rate of change in atmospheric CO_2 concentration across the same period.

Is this a surprise? Perhaps not. The concentration of CO_2 in the atmosphere is determined by interactions of four big processes (plus several smaller ones). These are:

- CO_2 emitted by burning fossil fuels (about 40,000 Mt CO_2 y^{-1}).
- CO_2 uptake by terrestrial vegetation (about 120,000 Mt CO_2 y^{-1}). However, much of the CO_2 taken up by vegetation plants is lost within a year or two (but it can take centuries for the wood in a tree trunk to be converted back to CO_2 via respiration of fungi and beetles (for example)), or through being burnt. The rate of uptake of CO_2 by vegetation is influenced by temperature, atmospheric CO_2 concentration and soil water availability and therefore changes month-by-month, year-by-year. There is abundant evidence that increasing global mean temperatures have increased the release of soil C to the atmosphere as soil respiration. Has this contributed to increased average annual atmospheric CO_2 concentration?
- The world's seas and oceans both absorb and release CO_2 from/to the atmosphere, depending on temperature, wind speed, atmospheric CO_2 concentrations and ocean currents, and uptake/release changes month-by-month, year-by-year. Cold water can absorb more CO_2 than warm water; currents that move water towards or away from the equator can result in that volume of water changing from being a sink (uptake of CO_2) to a source (releasing CO_2). Have increasing sea-surface temperatures reduced net uptake of CO_2

by oceans and contributed to increased average annual atmospheric CO_2 concentration? Net average annual marine uptake is between 5,000 and 14,000 Mt CO_2 y^{-1}.

- Land-use-change (converting the Amazon rainforest to cropland, for example) releases about 4,000 Mt CO_2 y^{-1}.

By analogy, if the concentration of CO_2 in the atmosphere is represented by a partially full bucket of water, we have several hands pouring different-sized cups of water daily into the bucket and several hands daily scooping out different volumes of water. Perhaps focusing solely on cutting fossil fuel emissions is not an effective strategy.

My questions to our political decision-makers are:

What evidence do they have that reducing Australia's emissions to net zero will reduce the rate of annual increase in atmospheric CO_2 concentrations?

What evidence do they have that reducing **global** annual emissions by (an unachievable) 50 % will prevent atmospheric CO_2 concentrations from continuing to rise?

A few statistics may also serve to highlight the low likelihood of success in significantly cutting global annual CO_2 emissions if the biggest emitters do not play ball:

1. The German Government has recently approved the expansion of the Garzweiler open cast coal-mine, which currently covers about 48 km^2 and produces lignite, a particularly polluting form of coal. Simultaneously, Germany is reopening five coal-fired power stations.
2. South Africa opened a new coal-fired power station in 2021 (the Medupi station) and opened another (the Kusile station) in 2022, with a combined capacity of about 30 million tonnes of coal burnt per year.
3. A 2022 report from the Centre for Research on Energy and Clean Air (CREA) and the Global Energy Monitor (GEM) demonstrated that approvals for coal-fired power plants, plus new construction plus new project announcements accelerated dramatically in China in 2022, with new

approvals reaching the highest level since 2015. **The coal-fired generating capacity starting construction in *China is six times larger than the sum of the rest of the world.*** This equates to two large coal-fired plants per week being approved.

4. India is currently constructing 32,000 MW of new coal-fired electricity generating capacity, according to the Global Energy Monitor. Assuming a 50-year lifespan for such stations, India will be importing coal for many more decades. Demand for coal in Asia is surging (for example, 23 new coal-fired plants in Indonesia are under construction or planned). For context, the UK's power generation capacity is less than 100,000 MW. So, 32,000 MW is about 32 % of the total UK power generation capacity.

5. Between 2022 and 2025, 27,300 MW of new natural gas-fired capacity are scheduled to come online in the United States. While natural gas is less polluting than coal, it remains a significant source of atmospheric CO_2.

Noting all the above, I believe it is crucial that we critically examine (i) our headlong rush in the West to stop using fossil fuels; and (ii) the reasons for the lack of response in rate of increase in CO_2 concentration during the three recent periods of reduced global CO_2 emissions.

PROBLEMATIC FEEDBACK LOOPS: DECARBONISING IS NOT SMART AND PROBABLY ISN'T ACHIEVABLE

For the purpose of discussion, I raise a (probably naive) suggestion. I doubt that it is contentious to state that we are in the midst of *three major feedback loops*. The first is the feedback loop whereby a small initial increase in temperature causes soil respiration to increase, which further increases CO_2 concentrations, which enhances global warming, which ….. repeat, **and** fossil fuel emissions are acting, and have been acting for two centuries, to increase temperatures, thereby compounding the original problem. The second feedback loop is centered on the observation that CO_2 solubility in oceans declines with a warming of oceans and therefore warming oceans release CO_2 into the atmosphere, thereby contributing to further warming. The effect is the same no matter the mechanism. The third feedback loop is centered on an overall decline in the Earth's albedo (reflectance) arising from declining snow/ice cover. This means that less solar radiation is reflected back to space and more is absorbed by Earth as a whole, thereby contributing to warming.

What is contentious is the suggestion that the absence of a response in the rate of increase in CO_2 concentration across the three time periods discussed above have arisen because soil respiration, plus ocean warming, plus reduced surface albedo (the reflectivity of the Earth's surface) have swamped the signal generated by changes in rates of fossil fuel burning. These three periods demonstrate that decarbonising the West will not achieve the desired aims. Perhaps it might be worth examining the question: is decarbonising Western society by the cessation of fossil fuel-burning a rational response to global warming in today's modern, complex, dangerous, geopolitical world? Can we afford to decarbonise the West in splendid isolation?

Is there an alternative approach to decarbonising the West? Perhaps we should be looking to maximise the greatest good for the largest number of people in the world when assigning resources to tackle climate change and a myriad of other socio-environmental-economic issues. This is discussed in my final chapter, which also asks, and answers, the question: is the Science of climate change settled? Spoiler alert: it certainly is not.

THE END

CHAPTER 12

WHAT IS SCIENCE AND THE SCIENTIFIC METHOD? WHY IS THE SCIENCE OF CLIMATE CHANGE SO TOTALLY NOT SETTLED?

The phrase 'The Science is settled' is frequently used in the media (print and electronic) when discussing climate change. This means that no one is allowed to question the existence, and more importantly, the causes, of climate change, and no one is allowed to question the optimal strategies required in the future to either prevent further increases in temperature and atmospheric CO_2 concentration or learn to live with climate change. Because – **The Science Is Settled**.........

This chapter addresses four big questions. These are:

e) What is Science and the scientific method?

f) Can Science ever be settled?

g) Is there evidence that human-induced increases in temperature are having a catastrophic impact on extremes, including drought, tropical storms (cyclones), the Southern Oscillation Index and the number and areal extent of wildfires?

h) Is there an alternative to the net zero path currently being promulgated throughout the West?

That is certainly a comprehensive collection of topics!

Before I can address these questions, some definitions and discussion are needed so that we don't get too many sidebars defining key terms within the main text.

SOME KEY CONCEPTS

> Theory *versus* hypothesis *versus* a scientific law
> Inference, logic, deduction, and induction
> Falsification *versus* proof
> Correlation *versus* causation

WHAT IS THE DIFFERENCE BETWEEN A HYPOTHESIS, A THEORY, AND A SCIENTIFIC LAW?

A hypothesis (plural hypotheses) is **a precisely formulated, testable (see falsification below) statement of what a scientist (or anyone else) predicts will be the outcome of a future study**.

Formulation of a hypothesis *frequently* (but not always) precedes data collection. We may hypothesise anything that our mind can conceive. We could hypothesise the existence of elephants covered in yellow and red polka dots, because we can conceive of such a thing. After extensive data collection (observations of elephants) across the globe, in the wild and in zoos, we may, despite extensive research, and because no such animal has been observed, reject the hypothesis. However, we cannot **conclude** that there are no elephants with red and yellow polka dots in existence because we have not observed every elephant in the world. A validated observation of a single yellow and red polka dotted elephant would be sufficient to support the hypothesis; the complete absence of any such observations following *extensive* research, would be sufficient **to reject the hypothesis,** but with the words "there is a very low probability of elephants existing with yellow and red polka dots".

Rejecting the hypothesis is not the same as *proving* the hypothesis to be wrong.

Hypotheses can, however, also be formulated following collection and consideration of a preliminary, usually limited, set of data. Hypotheses can also be generated from the application of prior knowledge about a topic, even in the absence of any *new* relevant data. The purpose of a good hypothesis is to focus the mind on what new data are required to support (not prove), or refute, the hypothesis. A good hypothesis will be open to testing by experiment or additional collection of robust data (see the discussion of *falsification* below).

A theory is a statement of a principle (or principles) presented as an explanation of observational data. In contrast to a *hypothesis*, a *theory* is constructed *after* the collection of sufficient data for generalised rules to be constructed using these data. A theory requires a (sufficiently large) body of evidence that supports the theory. Theories are strongly predictive. Darwin's *theory* of evolution by natural selection is a theory, not a hypothesis, because there is a vast amount of data, going back to the mid-1850s, that support it. Over the past century it has generated many predictions that, upon testing, were supported by robust data. A theory is a definitive explanation of some aspect of the natural world, as currently understood, but with the ***very strong proviso*** *that theories are discarded when new data become available that reliably refute the theory.*

Two formal definitions of scientific theories

The United States National Academy of Sciences defines scientific theories as follows:

"The formal scientific definition of (the word) theory differs from the everyday meaning of the word. It refers to a comprehensive explanation of some aspect of nature that is supported by a large body of evidence. Many scientific theories are so well established that no new evidence is likely to alter them substantially. For example, no new evidence will demonstrate that Earth does not orbit around the Sun (heliocentric theory), or that living things are not made of cells (cell theory), that matter is not composed of atoms (atomic theory of matter), or that the surface of Earth is not divided into solid plates that have moved over geological timescales (the theory of plate tectonics). One of the most useful properties of scientific theories is that they can be used to make predictions about natural events or phenomena that have not yet been observed."

The American Association for the Advancement of Science defines scientific theories as:

"a well-substantiated explanation of some aspect of the natural world, based on a body of facts that have been repeatedly confirmed through

> observation and experiment. Such fact-supported theories are not "guesses" but reliable accounts of the real world. The theory of biological evolution is more than "just a theory". It is as factual an explanation of the universe as the atomic theory of matter or the germ theory of disease. Our understanding of gravity is still a work in progress. But the phenomenon of gravity, like evolution, is an accepted fact."

A scientific law is a statement derived from many repeated and repeatable observations of a physical system (for example, observing the motion of our solar system's planets, or objects dropped from our hand). It is the existence of many *repeated* and repeatable observations that underpins a scientific law. A scientific law is deemed to be applicable anywhere in the Universe and is usually considered to be a true reflection of some (specified) aspect of a physical system. An early example is Newton's Law of Universal Gravitation (LUG), which states that all masses (for example atoms, people, moons, planets, and suns) attract all other masses with a force that is proportional to the product of their masses ($M \times m$, where M and m are the masses of the two objects) and inversely proportional to the square of the distance ($1/d^2$, where d is the distance between them). Newton applied inductive reasoning to empirical observations to derive his LUG. Inductive reasoning is defined on the next page of this chapter. The LUG is both *descriptive* and *predictive*, and like so many laws in science, can be written in a mathematical form (see reference to mathematical models below). Newton's LUG is still used extensively today to guide spacecraft, for example, even though Einstein's theory of relativity invalidates some aspects of Newtonian physics (in particular, the LUG applies *only* to weak gravitational fields (therefore doesn't apply close to black holes) and to speeds that are not too close to the speed of light, and it doesn't apply in the sub-atomic world). This tells us that scientific laws are subject to change (for example, limiting the scope of the applicability of the law), revision, and refutation in the same way that theories are subject to change, revision, and refutation. Laws provide a description of a set of specified events (for example, the motion of planets around our Sun; the motion of an object dropped from our hand on the surface of Earth (or Moon) but they do not provide a mechanistic explanation of why or how these events occur. Thus, scientific laws are empirically-derived conclusions that are descriptive, predictive, and open to falsification (see below). Scientific laws are descriptions of empirical regularities.

INFERENCE, LOGIC, DEDUCTION, AND INDUCTION

Inferences are conclusions that have been made using evidence and the application of reasoning; reasoning is the systematic application of logic. Logical reasoning derives a conclusion in a rigorous way from an initial set of premises and can identify correct and incorrect arguments. Premises are declarative statements (for example, "a triangle has three sides"). Declarative statements can be true (triangles have 3 sides) or false (triangles have four sides). A set of premises is generally the starting point for a logical argument, where logic is used to derive a valid conclusion from the initial set of premises.

There are traditionally two types of inference: deductive and inductive.

Deductive reasoning (or deduction) starts off from an idea/premise (for example, all humans are mortal). Following the initial premise/statement, an observation is made (for example, Riley is a human). From the initial premise and the observation, a conclusion can be drawn (Riley is mortal). In deductive reasoning, if the initial premise is true, and the observation is correct, then the conclusion (if logically derived) must be true. To put it more generally, If A = B, and B = C, and if each premise is true, then the conclusion A = C must also be true. Deduction usually moves from a generality (all humans are mortal) to a specific (Riley is mortal). Deductive reasoning is generally precise and quantifiable.

Inductive reasoning (or induction) starts from a specific observation (data point) and ends with a generality. For example, "I randomly withdrew 10 coins from a large black bag full of coins and all were dollar coins" is a specific observation. From this I can inductively infer (induce) that the black bag *probably* only contains dollar coins (a general statement about the nature of all coins in the bag). This is my hypothesis derived from the pattern in the observations. Of course, it might be that there are 1000 coins in the bag and only one is a 20-cent coin, the remaining 999 coins are one-dollar coins. The probability of extracting the single 20-cent coin on any withdrawal is low (1/1000 in the first withdrawal, 1/999 in the second…etc) and the probability of pulling 10 one-dollar coins in a row is about

99 %. The general statement at the end ("the black bag *probably* contains only one-dollar coins") is the hypothesis than could explain the original specific pattern in the 10 observations. Inductive reasoning tends to be probabilistic and often can't be falsified.

Scientists use both inductive and deductive reasoning during their working week/careers. One comparative representative of inductive and deductive reasoning can be written as:

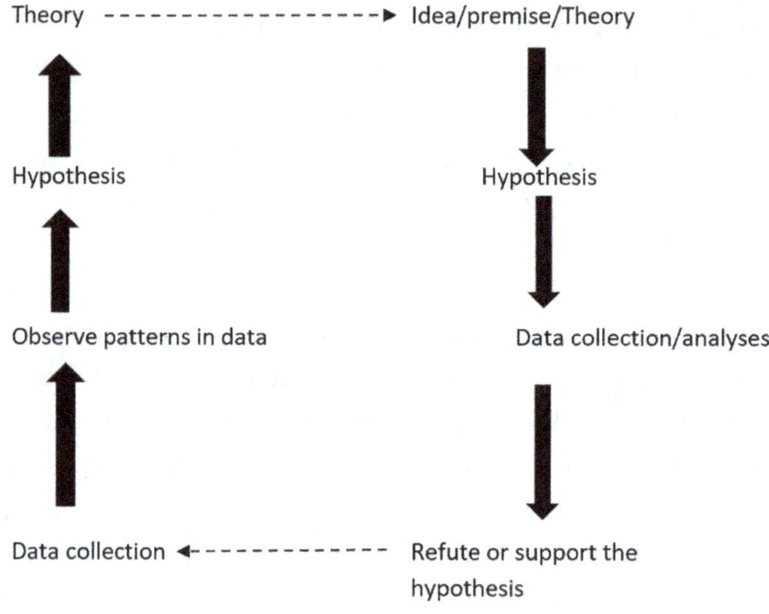

Figure 1: A comparative representation of inductive and deductive reasoning. It is possible to use both in any given research project, and the horizontal dashed arrows indicate the cyclic nature of the trajectory of many research projects.

From Figure 1 we can see that inductive reasoning can be used to develop hypotheses and theories and is characterised as moving from the specific (a specific and small data set or observation) to a broadly-based theory. In contrast, deductive reasoning is the opposite, and is used to test a hypothesis or theory. Many research projects may start with inductive reasoning, collect a small preliminary data set, produce a theory that leads to new hypotheses, which lead to new sets of data being collected, which are analysed to refute or support the hypotheses.

FALSIFICATION VERSUS PROOF

Karl Popper was an influential philosopher of Science throughout much of the 20th century. Perhaps his three biggest contributions were (i) identification of the problem of inductive reasoning in Science; (ii) the proposition that falsifiability was the key feature of Science; and (iii) that hypotheses and theories must be falsifiable to be accepted as valid hypotheses or theories. An alternative to the word falsifiable is "can be refuted". Put simply, a hypothesis or theory must be *predictive* and *testable* if it is to be deemed scientific (according to Karl Popper).

Philosophers of Science tell us that **the** problem with induction in Science is that we can't make an infinite number of observations. We could hypothesise that "all ravens are black". We can start testing this by observing a raven in the field. It is black. However, we cannot conclude that *all* ravens are black as this is a logical fallacy – we have not observed all ravens in the world. Thus, we cannot *prove* that all ravens are black without examining every raven in the world. Therefore, Popper argued that we cannot use induction to establish scientific laws and theories. Indeed, Popper believed that induction has no place in science. Sadly, he was wrong on these last two points. Philosophers of Science often don't inhabit the real world of scientific enquiry. I think we can safely and correctly apply inductive reasoning to conclude that there will be no observations in the future, that refute the theory that the planets of our solar system do indeed rotate around the Sun, and not *vice versa*. Similarly, we can safely and correctly apply inductive reasoning (Fig. 1) to conclude that when I drop a brick from my hand, and in the absence of an extreme updraft, the brick will fall to Earth in a direction that signifies the centre of Earth.

To overcome the problem of not being able to make an infinite number of observations (of the colour ravens, or the behaviour of apples dropping from a tree), Popper proposed that *falsifiability* is the means of moving from observations and experiments to theories and scientific laws. Specifically, Popper believed deductive reasoning was the basis of Science; valid hypotheses and theories are used to make predictions that are testable and any tests that show these predictions to be untrue result in the rejection of the hypothesis/theory. According to Popper, *falsifiability* is the *essential criterion* for distinguishing between Science and pseudo-science (the demarcation dispute: how to differentiate between Science, pseudo-science, and quackery). Finally, it must be emphasised that technically and logically, nothing can be proven in Science because we cannot make an infinite number of observations. However, we can disprove a hypothesis – for example, the hypothesis that *all* ravens are black would be disproved by an observation of a *single* non-black raven. Data can be said to *support* a hypothesis/theory, or can be said to *refute* or disprove a hypothesis/theory.

In the absence of an *infinite number of observations/tests*, data **cannot** prove a theory.

Now, it must be said that there are some who refute the Popperian view of the importance of deduction and falsifiability to Science. The identification of falsifiability as the means of separating Science from pseudo-Science and quackery has also been questioned. Thomas Kuhn was a prominent critic of Popper who believed that paradigm shifts (fundamental changes in foundational ideas and approaches within a scientific discipline) are central to distinguishing between Science and non-Science. On the first point (the importance of deduction and falsifiability in Science), I would say that Popper was right. On the second point (the identification of falsifiability as the means of separating Science from non-Science), he was right in most, but not all, instances. Astrology passes the Popperian criterion as being scientific because it is amenable to testing (that is, it is open to falsification), which means that by the Popperian standard, astrology would be counted as a legitimate Science, when it clearly is not. However, whenever it has been tested, it has failed the test and the hypothesis (the movements and relative positions of celestial bodies influences human behaviour and personalities) must be rejected. He was wrong about the unacceptability of inductive reasoning too (see above).

CORRELATION VERSUS CAUSATION

A correlation is a statistical statement about the existence, or not, of a relationship between two variables (things we can measure, such as age, height and weight). The two variables could be the age and height of a child.

In Figure 2 there is a highly significant exponential correlation between the size of the population of a small remote city in Western Australia (Perth), on the x-axis, and the annual average global price of gold, on the y-axis, both plotted for the period 1969 to 2022. We know the relationship is significant because the quoted p-value ($p < 0.001$) is much smaller than 0.05. In most science disciplines, a *p*-value smaller than 0.05 is taken to mean the relationship is significant. The 0.05 means that there is less than a 1 in 20 chance of the relationship arising by chance. While $p < 0.05$ is almost universally used as a statistically significant test result, I cannot find any reason published to substantiate why $p < 0.05$ is used and not, let us say, $p < 0.04$, or $p < 0.025$. The choice of a $p < 0.05$ threshold for statistical significance appears to be arbitrary. We also see that the coefficient of determination (R^2) is relatively close to 1 (the maximum value it can have is 1). An R^2 of 0.8 means that 80 % of the variation in the y-axis variable (the dependent variable; in this example, the average annual global price of gold) is explained by variation in the x-axis variable (the independent variable; in this example, the size of Perth's population).

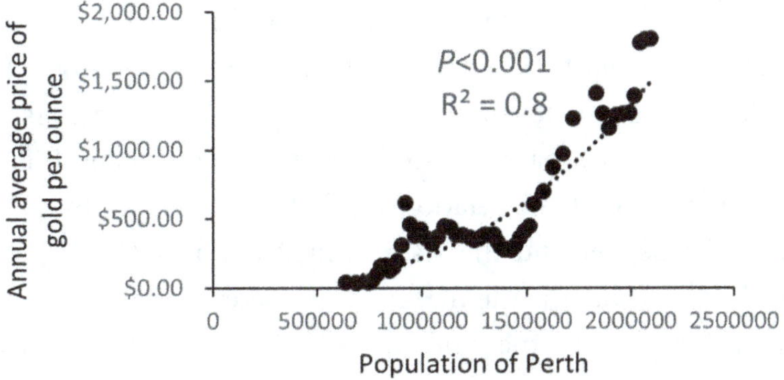

Figure 2: The relationship between population size of Perth, WA, and the average annual price of gold for the period 1969 to 2022 exhibits a highly significant (*p*-value much less than 0.05) exponential relationship (shown by the dotted line).

Does the strong statistical relationship between the two variables, as indicated by the small *p*-value and the large R^2, mean the increase in the size of the population of Perth caused the increase in average annual price of gold, over the period 1969 – 2022? Despite the very small *p*-value ($p<0.001$) and the large R^2 we are very unlikely to believe there is a *causal* relationship unless a plausible and testable hypothesis can be proposed to establish a mechanism by which the size of the population of a very remote city, Perth, can influence global annual gold prices. Do all (or a large fraction) of the residents partake in mining and produce a large fraction of the world's gold output? No, they do not. Western Australia (where Perth is located) produces less than 0.1 % of the global output of gold. Do they consume a very large fraction of global gold output and thereby affect its price? No, they do not. We must conclude that *the correlation is significant* but **attach no importance to this fact.**

In Figure 3 I have plotted the weight of a beaker sitting on an electronic balance as more and more water (of known and accurately measured volumes) are added to the beaker. The *p*-value is again very small (*p*<0.001) and the relationship, in this case linear, is highly significant. The R^2 is equal to 1, meaning *all* the change in weight is accounted for by the change in the volume (and hence weight) of water added to the beaker. Is this a causal relationship? Yes, it is, but this conclusion is reached not because R^2 is equal to 1 or the *p*-value is again much smaller than 0.05 (meaning a very statistically significant relationship). We conclude it is a causal relationship because we have a clear hypothesis as to why this is causal (that is, the addition of water explains the increase in weight of the beaker plus water) and why R^2 is equal to 1. The plausibility of this causality is supported by other information. Thus, the density of water is 1 g cm^{-3}. This means that for every 1 cm^3 of water added, 1 g of mass is added to the beaker + water. It is a mathematical certainty, therefore, that the relationship between the volume of water added and the mass of the beaker and the water added is linear (Fig. 3). Note that the mass of the beaker plus water is the mass of the beaker plus the sum of all volumes of water added prior, plus the mass of the addition of a new volume of water. We can, from Figure 3, also determine the mass of the empty beaker: it is 100 g.

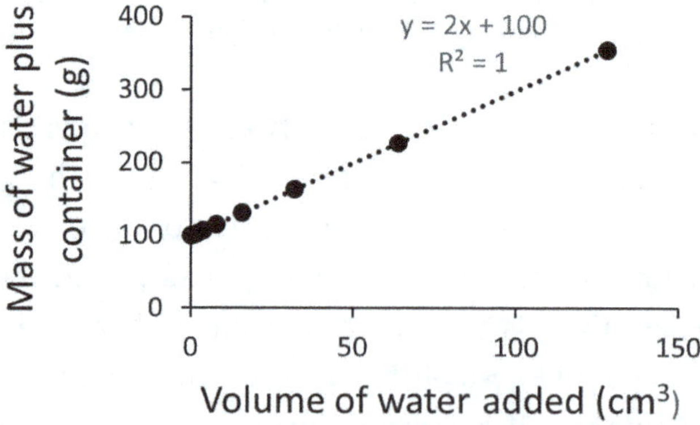

Figure 3: The relationship between the volume of water added to beaker and the mass of the beaker plus water added exhibits a highly significant linear relationship (shown by the dashed line). Note that the total volume of water added at each point is the sum of all preceding volumes added plus the latest volume added. *p* < 0.001.

The *key message* here is that having a highly significant regression (linear or otherwise) and a large R^2 does not tell us whether the relationship is a mere correlation, with no mechanistic link between the two variables, or is a causal relationship, implying a link between the two variables (changes in the x-axis variable causes a mechanistically plausible and *predictable change* in the value of the y-axis variable).

SCIENCE, SCIENTIFIC PLURALISM AND THE SCIENTIFIC METHOD?

Science is first and foremost *a process or set of processes*. Some definitions (see Text box below), but not all, include a second clause, namely, that the body of knowledge accumulated through the application of the Scientific method (more of this below) is also Science (it isn't). Like the Stanford Encyclopedia of Philosophy, I don't subscribe to the view that a body of knowledge can be labelled Science. A body of knowledge may be scientifically obtained, but Science is the process by which new knowledge is obtained. It might not be the only way new knowledge can be obtained, but it is an exceedingly useful, robust, and productive way of acquiring new knowledge. But a body of knowledge *per se* does not constitute Science.

Four formal definitions of Science

"both a body of knowledge and a process. Science is a process of discovery that allows us to link isolated facts into coherent and comprehensive understandings of the natural world"

— The University of California, Berkeley

"the pursuit of knowledge and the understanding of our environment following a systematic methodology based on evidence. It can be defined as both a body of knowledge we already have and the pursuit of a deeper understanding of both ourselves and the world around us"

— The Royal Societies of Australia

"both a body of knowledge (the things we have already discovered), and the process of acquiring new knowledge (through observation and experimentation—testing and hypothesising). Both knowledge and process are interdependent, since the knowledge acquired depends on the questions asked and the methods used to find the answers"

— The Australian Academy of Sciences

"the pursuit and application of knowledge and understanding of the natural and social world following a systematic methodology based on evidence"

— The Science Council

This, of course, begs the question – what is the process? The process of scientific enquiry is almost universally referred to as the Scientific Method (SM), so I shall stick with that name.

A typical representation of the SM, from the Khan Academy and other educational fora, including Universities, is given below:

1. Make an observation.
2. Ask a question pertaining to that observation.
3. Form a hypothesis (a testable explanation).
4. Make a prediction based on the hypothesis.
5. Test the prediction experimentally.
6. Repeat and use the accumulating results to make new hypotheses or predictions. Return to (5) and then (6).

NASA lists the following steps for the SM:

1. Ask a question or make a statement that you can test by an experiment. This statement is called a **hypothesis.** The hypothesis defines the purpose of your experiment.
2. Define the parts of your experiment that will change. These are called variables.
3. Find out what people have already said or written about the subject. Has anyone else done the same experiment? Even so, you can still repeat their experiment to see if you get the same result.

4. Perform the experiment.

5. Analyse the data.

6. Draw conclusions and decide whether the data support or refute the hypothesis as stated.

These definitions probably encapsulate what the public, probably all schools, and indeed many new or poorly trained scientists think is the process *universally* followed by scientists. An obvious problem with this description of the SM is that there is no mention of where a question might come from, nor the place of theoretical studies where experiments are not possible.

Science incorporates many disciplines and sub-disciplines. Just five minutes of thought generated this table (Table 1) of disciplines, and it is by no means exhaustive.

Table 1: Some of the many disciplines and sub-disciplines of Science.

Life sciences/ Biology	Chemistry	Physics	Astronomy	Atmospheric sciences
Sub-disciplines Physiology Taxonomy Ecology Forestry Agronomy Evolution Animal behaviour Virology Parasitology Botany Zoology	*Sub-disciplines* Inorganic Organic Physical Analytical Environmental Biochemical Chemical engineering	*Sub-disciplines* Classical Modern Nuclear Atomic Thermodynamics Optics Geophysics Mechanics Biophysics	*Sub-disciplines* Astrophysics Cosmology Astrobiology Astrogeology Asteroseismology	*Sub-disciplines* Meteorology Micrometeorology Climatology Atmospheric physics Atmospheric chemistry Hydrometeorology

Is it likely that a single method can be applied across these disciplines/sub-disciplines? Nobel Laureate Weinberg said, in 1995:

"The fact that the standards of scientific success shift with time does not only make the philosophy of science difficult; it also raises problems for the public understanding of science. *We do not have a fixed scientific method to rally around and defend*". (my emphasis added).

Here is a distillation of my view of the SM, gleaned after 40 years in the game.

There are multiple steps and options in applying the SM to a particular problem/question/experiment. These can be summarised thus:

Step 1:
Option 1) Read the scientific literature relating to a specific narrow area of interest and identify (a) a knowledge gap in the literature of this topic; or (b) a question raised in the literature that remains unanswered; or (c) a data set that can be interpreted in a new and as-yet unexamined way.
OR:
Option 2) Conduct a thought experiment entirely independent of any literature or data sets (for example, in 1895 probably no-one had asked the question: "what would the Universe look like if I were travelling on a photon?" until Einstein asked it a few years later). Thought experiments require imagination, creativity, and intelligence, and they are a valid process within the SM. They often address theoretical problems and use mathematics to answer them. A similar approach uses the question: "what would happen if...?" but such questions are generally open to experiment. The question "what would happen if I measured the resistance of a length of copper wire at room temperature and then at the temperature of liquid nitrogen?", can be addressed experimentally. The question "what would the Universe look like if I travelled at the speed of light?" cannot be experimentally addressed.
OR:
Option 3) Collect a set of data on, let us say, the distribution of barnacles on rocks in a rocky bay in south Wales, without any underlying hypotheses in mind and without any significant or deep knowledge of the literature on this topic. Measure the distribution of barnacles (numbers of barnacles per m^2 of rock surface) and a whole range of other variables that we think might be important (for example, N, S, E and W facing surfaces; height above low tide; annual and seasonal average minimum and maximum temperatures of air and seawater at the site; distance from the high tide mark) and then do a series of increasingly sophisticated statistical tests among the dependent variable (barnacle distribution) and independent variables (N, S, W, E facing aspects, temperature, etc). Find significant correlations and *then think some more to define some questions and then hypotheses for future testing*. This is not an approach that is rec-

ommended because it lacks an underlying hypothesis, lacks any significant or deep knowledge of the literature, and often results in poor experimental design. However, this approach has been, and is, used by poorly trained, novice, or lazy scientists. But it can yield useful results, hypotheses, and theories.

OR:

Option 4) The advent of widespread high-speed computing and parallel computing has introduced a radically different approach to doing Science to many disciplines. Advanced mathematical modelling was once limited to relatively few researchers because of the time taken to crunch, by hand, the numbers required, and the relatively few scientists with the requisite mathematical skills to undertake such modelling. This is no longer true. One useful approach to computer modelling is to start with a single, relatively simple, and relevant process for which the mathematical model is known (for example, soil infiltration into a sandy soil if one is interested in catchment hydrology), generate the computer code required to run the mathematics, and compare the model output with experimental lab and field data. It is then possible to add more and more processes (run-off of water down a slope; evaporation of water from a wet soil surface or a wet canopy; transpiration of water through a plant canopy as a function of soil water content; rainfall-river-flow relationships; deep percolation of water to groundwater) by coupling each process to all other relevant processes (linked subroutines) and generate a single computer model of catchment hydrology. At each stage of development of each sub-process, comparisons are made between model outputs and field and lab data. The development of such computer models requires an understanding of applied physics (including soil and atmospheric physics), applied mathematics, plant ecophysiology and micrometeorology and computer coding. It generally takes several years/decades of development before any given computer model becomes accepted as a reasonable representation of a set of interacting processes. These models provide new insights and allow new methods of testing our understanding of a set of interacting processes that probably would be beyond our comprehension in the absence of such models within fast computers. Of course, the jeopardy is that too much reliance can be made on the interactions and outputs of extremely complex models (see Can Science Ever Be Settled, below).

Step 2:

Option 1) After formulating a question *either* (a) design a rigorous experiment to answer the question; *or* (b) develop the question into a testable hypothesis and then design a rigorous experiment to test the hypothesis. A rigorous experiment in experimental sciences includes replication (doing the experiment several times, usually, but not always, simultaneously), a control (that is, no treatment applied) and an experimental (treatments are applied) set of observations (which means, for example, in a test of the hypothesis that increased supply of soil nitrogen increases crop yield, several plots of plants receive one of several rates of addition of nitrogen, and several plots act as the control treatment, where no nitrogen is added). Note that there is a vast number of perfectly acceptable journal papers, published in reputable journals, that seek to answer questions rather than seek to falsify (i.e., not prove) hypotheses.

OR:

Option 2): Some questions/hypotheses/sub-disciplines of science are either not amenable, to experimental testing and/or replication or the application of treatments/controls. The hypothesis "there is life on other planets outside of our solar system" is not amenable to an experimental test (yet). It might be amenable to a vast exercise of observation of the Universe, but it isn't amenable to an experimental test. Similarly, the answer to the question "how did life originate on Earth" remains unknown. It is true that multiple experiments in various laboratories have zapped various soups, composed of different mixtures of simple organic and inorganic molecules, with electrical discharges, UV radiation and X-rays, and have been able to produce some of the simpler molecules that we find in living cells, but the *answer* to the *question* remains a mystery. So, what makes questions such as these scientific? First, the observations made are repeatable by multiple laboratories around the world. Second, the sensors and equipment used are rigorously calibrated, for accuracy and precision and purity of measurement (they measure only what we think they measure). Finally, hypotheses derived, for example, from the question "how did life originate on Earth?", can be tested/falsified experimentally. We can test the hypothesis that spark discharges or UV radiation above soups of various very simple organic and inorganic elements and compounds can produce new, more complex, compounds not initially present and that some of these compounds are indeed found in living cells. However, testing these hypotheses does not address the primary question: how did life originate on Earth? This is essentially because we can't go back in time and take a look. The take home message is that the SM does not always include replication or experimental falsification of hypotheses.

Step 3: Analyse the results of the experiment statistically and determine whether the results answer the question adequately and rigorously, or whether the results support or refute the hypothesis.

In conclusion, we must, I think, accept the proposition of *"Scientific Pluralism"*. Scientific pluralism maintains that there is no single universal SM. Pluralists note that across scientific disciplines, different SM and theoretical models are used to great effect and that those who believe there is a single, unified, SM, need to categorically demonstrate this. To-date, this has not been achieved (and I doubt it will be).

CAN SCIENCE EVER BE SETTLED?

Science is a process, or a number of processes. The question "can Science ever be settled?" really means "can the process of Scientific enquiry stop in relation to a specific problem/question/ hypothesis/theory/law?"

In the real world of practical Science, there are some highly specific cases where the answer to this question is **yes**. No one in a Science lab somewhere is mapping the movement of planets to test the Heliocentric Law of our solar system and no-one is calculating the value of G in Newton's Law of Universal Gravity (G = 6.67430×10^{-11} Newtons kg^{-2} m^2). Similarly, no one is looking for a life form that does not conform to the cellular theory of life (namely that all living things are made of one-or-more cells; this is why prions aren't alive (Chapter 7)). So in some specific and limited cases, we can say "The Science is Settled".

But, note that these are highly specific, and very old, topics (old in the sense that they were answered more than one century ago, or, if you prefer, they have not been falsified for more than a century).

Despite what we routinely read or hear in the popular media, it is generally acknowledged, outside of mainstream media, that science is almost never 'settled'. Put the words 'can science ever be settled' into Google and the answer is a resounding **NO**. Science is rarely settled because, while explanations for things can be well understood today, there is always more to know, and, most importantly perhaps, scientific consensus has frequently been wrong (Earth is not the centre of the Universe; Earth is not a few thousand years old; living organisms (e.g., maggots) do not arise from non-living matter (rotting meat)).

What of the question: is the Science of climate change settled? Many will tell you it is settled. For example, visit https://thelogicofscience.com/2015/08/16/settled-science-part-1-is-science-ever-actually-settled/ But it ain't. I will show why shortly, but first, what is climate change? Simply put, climate change is a change in the long-term (minimum 30-year average) of a number of climate variables, considered singly or grouped together (Chapter 11). Relevant climate variables include rainfall, surface temperatures, extent and severity of droughts, humidity and cyclones, for example. While not a climate variable, wildfires are often grouped in discussions of climate change because these are affected by drought, humidity, and surface temperatures. Large-scale natural cycles in weather patterns (for example, the El Niño–Southern Oscillation) can, over a suitable time-frame, be considered within Climate Change Science.

But here is why the Science is not settled......

The simplest demonstration of the fact that the Science of climate change is not settled is given in Figure 4 which shows an approximately 2.5 °C difference in the predicted temperature increase according to eight different climate models. The 'shakiness' of these models is further illustrated by examining model predictions with actual data. For example, the Intergovernmental Panel on Climate Change (IPCC) fifth Assessment Report stated that the observed rate of climate warming for the period 1998 – 2012 was lower than that predicted by 111 of 114 GCMs (General Circulation Models).

There are many other examples of the considerable uncertainties and biases in global climate models.

In 2022, Wei and colleagues compared 23 GCMs from phase 6 (a recent iteration) of the IPCC Coupled Multi-Model Intercomparison project and concluded that the outputs of the multi-model ensemble significantly overestimated annual total rainfall in the Yangtze River Basin (the third longest river globally; it drains

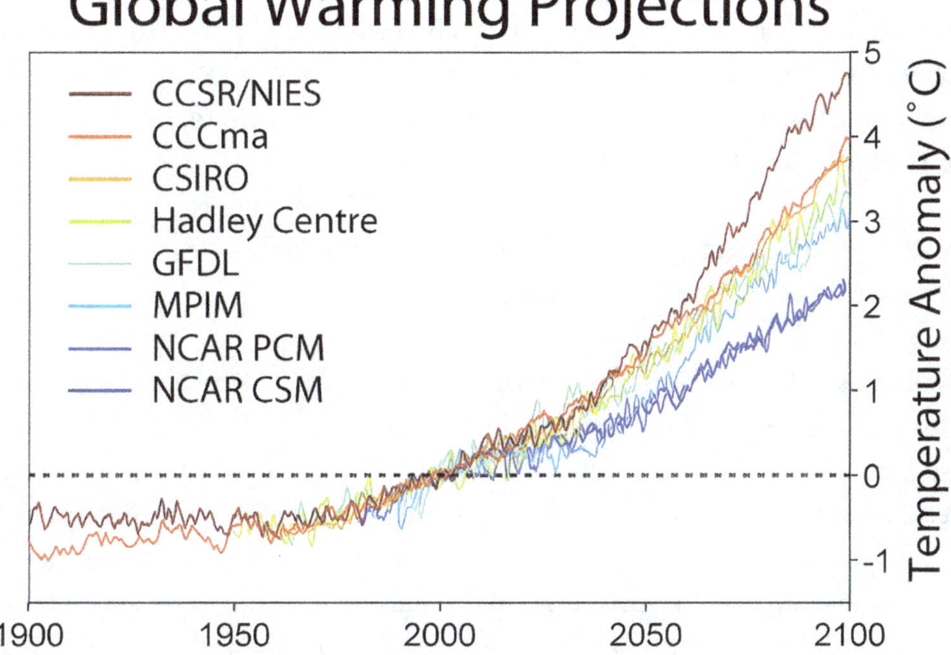

Figure 4: Temperature predictions of eight GCMs. Note the range of predictions for year 2100. Predicted temperature increases range from about 2.2 °C to about 4.8 °C by 2100. From: https://en.wikipedia.org/wiki/General_circulation_model

20 % of China and accounts for about 20 % of China's GDP). The IPCC Coupled Multi-Model Intercomparison project combined the outputs of many GCMS and an average response was calculated for, let us say, annual rainfall. This was done in an attempt to minimise the effects of model errors and biases by, it is hoped, errors and biases of different models cancelling out across multiple models. They concluded:

"Extreme precipitation is highly parameterization-dependent; the convection schemes in CMIP6 *are imperfect*, which cause *relatively large biases* from observations and *further lead to uncertainties in projections*" (Wei et al., 2022, International Journal of Climatology).

Also in 2022, Oluwafemi and colleagues compared the global and continental outputs of 14 climate indices produced by 11 GCMs and concluded:

"individual GCM results often varied significantly between regions and models and *there were observable biases in the climatology of various indices.* These inconsistencies could be attributed to various land surface schemes and simulations of features such as vegetation and orography..." (Oluwafemi et al., 2022, Scientific Reports).

In 2013, Collins and colleagues highlighted the inadequacies in climate modelling and therefore climate predictions and wrote:

"...many small-scale physical, biological and chemical processes, including cloud processes, cannot be described by those equations, either because we lack the computational ability to describe the system at a fine-enough resolution to directly simulate these processes, or because we still have only a partial understanding of the mechanisms driving these processes. Those need, instead, to be **approximated** by parameterisations within climate models, through which a mathematical relation between directly simulated and approximated quantities is established, often on the basis of observed behaviour. There are **various alternative and equally plausible numerical representations**, solutions and approximations for modelling the climate system, given the limitations in computing and observations. This .... results in a range of plausible climate change projections at global and regional scales. This range provides a basis for quantifying uncertainty in the projections, but because the number of models is relatively small, and the contribution of model output to public archives is voluntary, the sampling of possible futures is **neither systematic nor comprehensive**. Also, **some inadequacies persist that are common to all models**; different models have different strengths and weaknesses; it is not yet clear which aspects of the quality of the simulations that can be evaluated through observations should guide our evaluation of future model simulations." (My emphasis added).

Let me make it clear that I am not attempting to cast aspersions on GCMs, nor question their usefulness or importance. I am merely addressing the question: is the Science of climate change settled? And it isn't.

This does raise the question: what exactly is "The Science of climate change"?

Science is, as I have demonstrated, a process, or a series of linked processes. The meaning of the phrase "The Science of climate change" is captured in the following figure:

Hypothesis 1: Atmospheric CO_2 concentrations have been increasing since the Industrial Revolution.

PLUS

Hypothesis 2: Increased atmospheric CO_2 concentrations that have occurred since the Industrial Revolution have caused global/continental/regional-scale climate change in the 20th and 21st centuries.

PLUS

Hypothesis 3: Stopping the burning of fossil fuels is the optimal solution to the problem of climate change and is rational, economically feasible, and feasible from both an engineering and geopolitical strategic viewpoint.

Figure 5: The Science of climate change as represented in three hypotheses.

That atmospheric CO_2 concentrations have been increasing since the Industrial Revolution is beyond doubt. The following figure (Fig. 6) is taken from Chapter 11, which discusses the difference between weather and climate and demonstrates the anthropogenic sources of this increase in atmospheric concentration of CO_2. There is irrefutable evidence to support hypothesis 1 in Figure 5.

No-one who understands the meaning of the word climate can state that climate is *not* changing. Climates (Earth doesn't have one climate, it has very many, including polar, temperate, Mediterranean, tropical, and arid...... to name just a few) have been changing for ever (Chapter 3). The Huronian glaciation occurred between 2.45 and 2.22 Gya (billion years ago) and covered entire continents and extended from polar to low latitudes. Before that, Earth's climate was much warmer and more humid, and after that, it was also warmer and more humid. Four additional major ice ages are also recognised (Cryogenian (720 – 625 Mya); Andean-Sa-

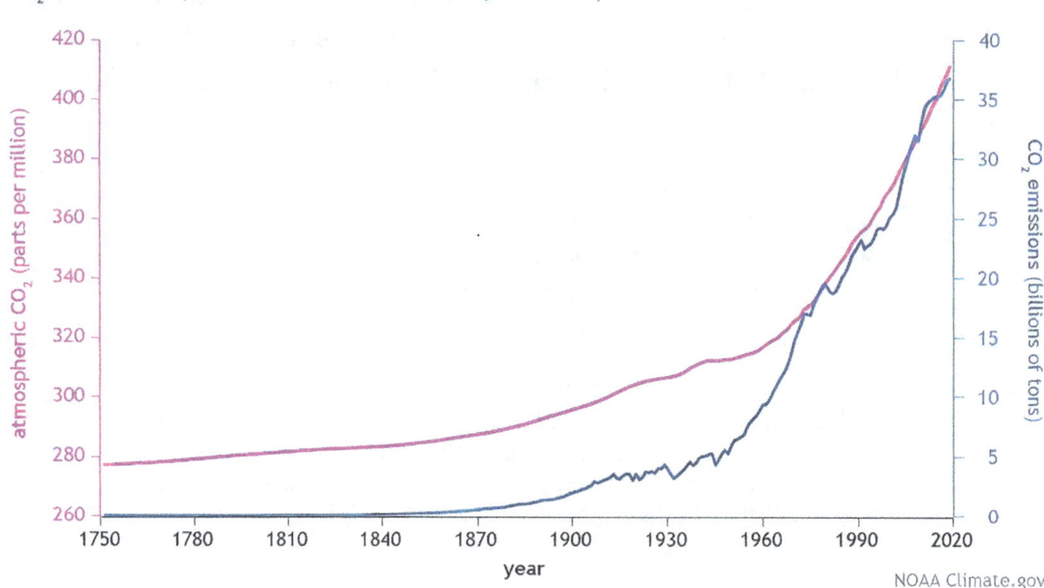

Figure 6: Atmospheric CO_2 concentrations since the Industrial Revolution. Data from NOAA and ETHZ. CO_2 emissions data from Our World in Data and the Global Carbon Project. Source: https://www.climate.gov/news-features/understanding-climate/climate-change-atmospheric-carbon-dioxide

haran (460 – 420 Mya); late Paleozoic (360 – 255 Mya); and the Quaternary (2.58 Mya) to the present day (we have polar ice caps and glaciers) ice age. The Little Ice Age affected Europe and the USA in the 14th to 19th century (hence Dickens having a white Christmas in all his novels) and the River Thames froze over in London to a considerable depth in winter between 1600 and 1814. Conversely, the Medieval Warm Period occurred across the north Atlantic (including northern Europe) between about 950 and 1250 AD (but was not a global phenomenon). Along with changes in the areal extent of ice, regional and continental temperature, and surface albedo, there were also changes in rainfall, humidity, river flows, and ocean currents. Climate has been changing for billions of years. To be a climate change sceptic, one must not understand what climate is and what it has been doing for the past 4 billion years. **However**, establishing the causal link between changes in atmospheric CO_2 concentrations and changes in climate is a little trickier. How, exactly is this being attempted, and is this "all settled"?

HOW DO CLIMATOLOGISTS INVESTIGATE THE LINK BETWEEN ATMOSPHERIC CHEMISTRY AND REGIONAL/CONTINENTAL/GLOBAL CLIMATES?

Climatology is heavily reliant on computer models. Massive models, called *Global Circulation Models (GCMs)* or *General Circulation Models (GCMs)* or *Global Climate Models (GCMs)* underpin climate change science. To run such huge and complex (see below) models requires massive computing grunt, using supercomputers capable of doing billions of calculations per second. They generate an awful lot of heat too (the computers, I mean).

WHY ARE GCMS SO LARGE AND COMPLEX?

Imagine Earth to be a simple sphere. Imagine drawing a grid on the surface of Earth (including the oceans and seas). The surface area of Earth is about 510,000,000 km^2. If each grid square is 100 x 100 km (total area = 10,000 km^2) there will be 51,000 squares covering the surface of Earth. Now, stack the grids vertically so there are 30 layers. We now have 1.53 million 'cubes' (not really cubes, but distorted cubes, called grid cells). We now have atmospheric cubes, oceanic cubes (these sit directly above oceans), terrestrial cubes (which sit directly above terrestrial surfaces and ice cubes located above ice. Figure 7 is a representation of these "cubes".

For each cube, the land and oceanic surface characteristics (for example, vegetation type, topography (slope, aspect, altitude)) and temperature (terrestrial and oceanic), oceanic salinity and currents and mixing, and atmospheric temperature and pressure, must be calculated. To make it a little harder, the rotation of Earth must be considered because this affects the transfer of heat and momentum between adjacent grid cells. Other variables need defining for each cell too, including incoming solar radiation levels (which vary sub-hourly, hourly, seasonally, and inter-annually), cloud cover, albedo (surface reflectance). Now the models need to compute all of these variables and processes as a function of time (half-hourly time steps are often used) and then include all interactions among adjacent grid cells. For example, atmospheric temperature and humidity, and wind speed and direction, affect the formation of clouds, which affect the albedo of a grid cell that now contains a cloud, and the change in albedo affects the temperature of that grid

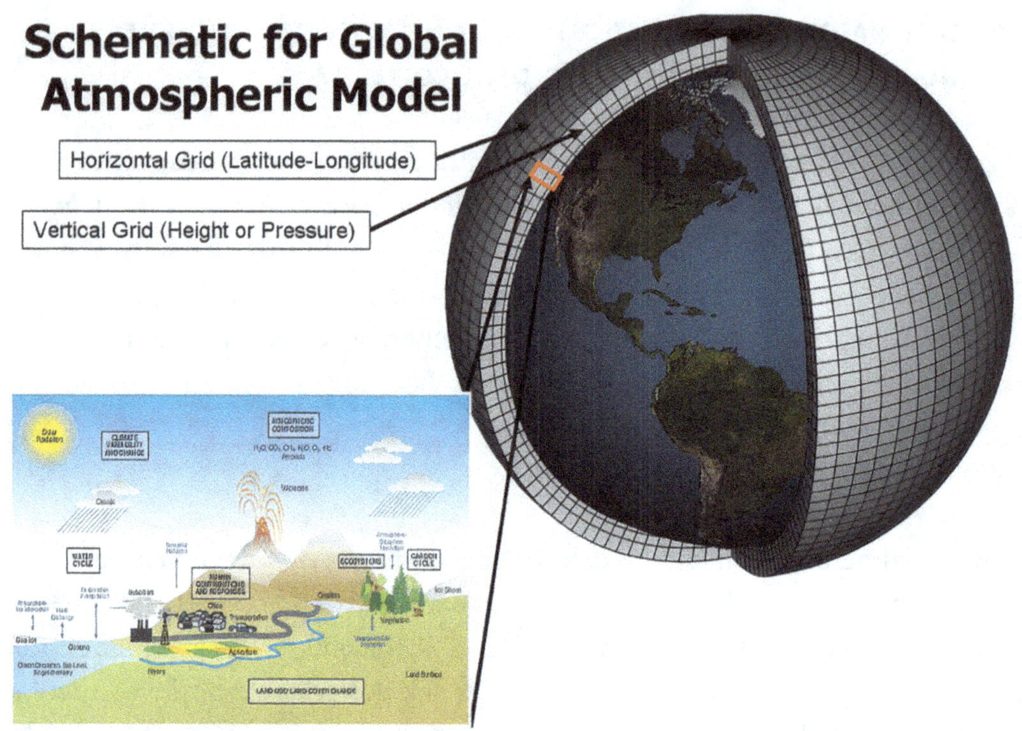

Figure 7: A schematic diagram of the representation of grid boxes on and above the surface of Earth used in a global atmospheric model. From: https://www.gfdl.noaa.gov/climate-modeling/ (NOAA figures are provided free of copyright).

cell but also the absorption of heat (infrared) emitted by the ground in that grid cell...... you get the picture.

Simulations within GCMs include the formation and movement of tropical cyclones, for example. Simulations obey all fundamental laws of physics (for example, energy and mass cannot be destroyed or created). Some processes are not simulated, they are *parameterised* because they occur at too small a spatial-scale for them to be included in a GCM. The formation and composition of aerosols, for example, is a parameterised process relying on field data (often derived from satellite observations). Similarly, convection and cloud formation are parameterised processes, too.

Most recently, GCMs include links to Earth System Models (ESMs) which include marine and terrestrial biogeochemical processes, vegetation-carbon interactions (uptake of CO_2 through photosynthesis and release of CO_2 via respiration) and atmospheric and upper ocean chemistry. Rainfall, surface flows of water, and vegetation responses to rainfall are included in coupled models (coupling of GCMs and ESMs).

Is there only one GCM currently in use? Certainly not. I think there are at least a couple of dozen GCMs currently in use. Do they all agree with each other – certainly not, otherwise there would be little point in having so many. And herein lies one reason why "Climate Science is not settled…..". The different models developed and run by, for example, the UK Met Office, the US National Oceanic and Atmospheric Administration Geophysics Fluid Dynamics Lab, and the CSIRO in Australia, have their own GCMs. While the Laws of Physics are the same in each GCM, the exact formulation of parameters, for example, differ slightly, leading to divergence in the predictions generated for future climates. The other reason for the Science not being Settled is because of ignorance on our part. We are not yet able to mathematically describe all of the physics that we need – for example, convection, aerosol formation, cloud formation and the formation of the Intertropical Convergence Zone around the equator remain especially problematic and until these can be reliably, accurately and precisely incorporated into coupled GCMs, more science is needed. Currently, values assigned to radiative forcings (see text box below) differ significantly among GCMS, another source of variation in model predictions for the future.

Radiative Forcings

The main inputs to a GCM are a range of variables that determine the fraction of incoming solar energy absorbed by the Earth's surface (terrestrial and oceanic) and atmosphere. These variables include the amount of solar radiation arriving at the top of the Earth's atmosphere (which varies seasonally and as a function of sunspot activity), the concentration of greenhouse gases, the concentration of aerosols in the atmosphere, which reflect, absorb, and scatter radiation and influence cloud formation, and the total area of the Earth's surface covered by ice (which affects albedo). These variables are known as *forcings*. According to the National Oceanic and Atmospheric Administration, forcings are the "prime movers of climate change and the main differences among GCMs involve aerosols. The impact of aerosols on the Earth's energy balance remains unclear".

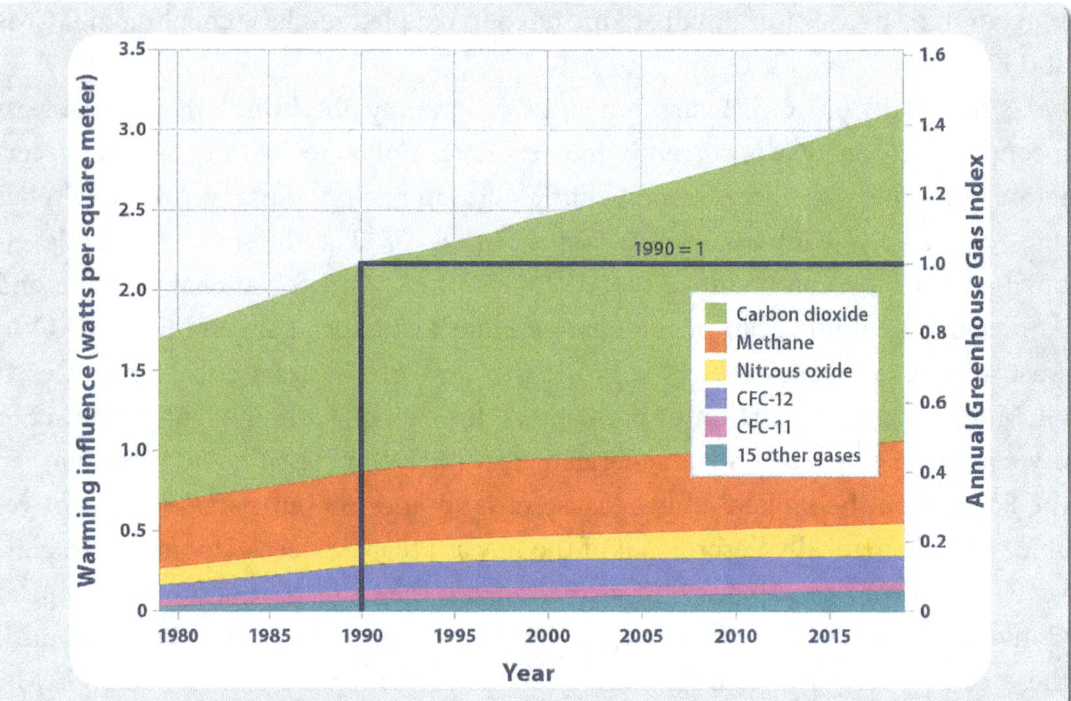

Figure 8: Radiative Forcing Caused by Major Long-Lived Greenhouse Gases, 1979-2019. From: https://www.epa.gov/climate-indicators/climate-change-indicators-climate-forcing

The amount of radiative forcing caused by various greenhouse gases has changed since the Industrial Revolution because of changes in concentration of these gases. Radiative forcing is calculated in watts per square metre, which represents the size of the energy imbalance in the atmosphere. A positive forcing means a warming of Earth. On the right side of the graph, radiative forcing has been converted to the Annual Greenhouse Gas Index, which is set to a value of 1.0 for 1990.

Clearly, the Science of climate change is not settled…..

A question that no-one ever asks explicitly is: why have we been burning more and more coal, oil, and gas for the past 1000 years, but especially for the past 250 years? Figure 9 provides a very strong correlation between human population size and atmospheric CO_2 concentrations for the past 100 years. The R^2 is almost 1. Is this relationship causal? I would argue that it must be.

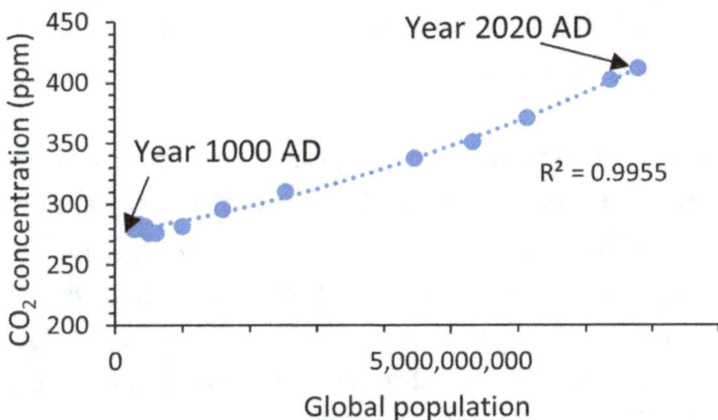

Figure 9: Atmospheric CO_2 concentrations have increased as global population numbers have increased. The coefficient of determination is very close to 1, indicating that almost all (99.55 %) of the variation in atmospheric CO_2 concentration can be explained by increased population size.

Why do I argue it must be causal? Several lines of reasoning:

a) as population numbers increase, even if energy consumption per person remains static, total global energy consumption (and hence CO_2 release) increases;

b) as population numbers increase, energy input to the means of production and distribution increase, including the synthesis of fertilisers, pesticides, herbicides, and fossil fuel-driven storage (often chilled or in tin/aluminium cans) and transport of food, regionally, continentally, and inter-continentally also increases;

c) as population numbers increase, energy consumption for heating/ cooling, construction of homes and other infrastructure (roads, schools, hospitals, industrial infrastructure), increase. Note, for example, that the production of 1000 tonnes of cement releases about 1 tonne of CO_2. Ever-increasing speeds of transport for increasing numbers of people result in increased energy consumption in absolute terms and per person;

d) as population numbers increase, complexity and expectations across the developed world, in terms of communications, computing power, and life-style associated energy consumption (especially disposable/discretionary income spent on leisure activities and average distance travelled per person in a lifetime) increase in absolute terms, and also in rates of consumption/ expenditure per person.

Put all these factors together, and a myriad of others, including deforestation (for food production and housing) and the spread of industrialisation, and we can safely conclude that global energy expenditure increased with increased global population. The **benefits** *of such increased expenditure of energy per person* can most simply be expressed as *increased life expectancy* observed over the past 250 years (Fig. 10). It is no accident that the dramatic increase in life expectancy for most of the world (excluding Africa and Asia) started at, or shortly after, the Industrial Revolution. Africa and Asia experienced the benefits of the Industrial Revolution later than the Americas and Europe.

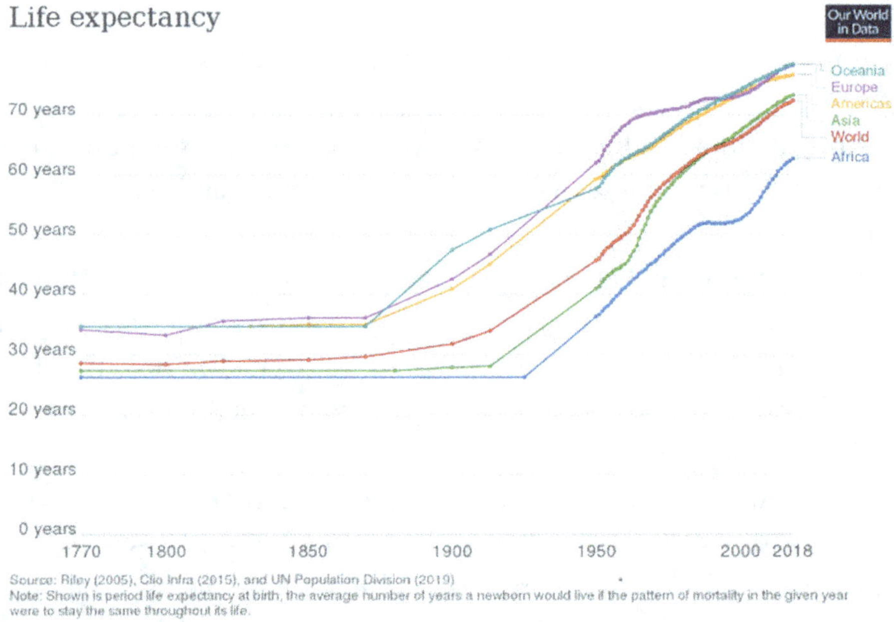

Figure 10: Average life expectancy was brutally short for most of the history of *H. sapiens*. It was the arrival of the Industrial Revolution that dramatically increased average life expectancy, through several mechanisms, most of which can be understood in terms of increased societal wealth.
From: https://en.wikipedia.org/wiki/Life_expectancy#/media/File:Life_expectancy_ by_world_region,_ from_1770_to_2018.svg. This image is licensed under the Creative Commons Attribution 4.0 International license. Image produced by Max Roser, data from https://ourworldindata.org/life-expectancy

The importance of the correlation between lifespan and CO_2 concentration (Figs. 9, 10) is discussed later in relation to hypothesis 3; more specifically, in relation to the moral argument put forward by many developing nations.

HYPOTHESES 2 AND 3 FROM FIGURE 5

What of hypotheses 2 and 3 of Figure 5? Hypothesis 2 is that global/continental/ regional-scale climates and weather extremes are changing in agreement with hypothesis 1. There are three key features of climates that are most relevant here (from a regional and global human perspective). The first is temperature (by which I mean the temperatures we (humans) experience on the surface of Earth); the second is drought; and the third is tropical cyclones. However, I shall also include a geographically large weather cycle that affects a large fraction of the globe, including Australia, the tropical Pacific, the western coast of South America, India, and Indonesia. This is the **El Niño–Southern Oscillation (ENSO)**. I choose this because of its integration of sea-surface temperatures, wind patterns, rainfall and atmospheric temperatures. If anything is likely to reveal changes in climate over the past 120 years, ENSO must. Furthermore, I include a brief consideration of wildfire data from several sources, because of the recent pronounced and extreme prevalence of wildfires in the past 6 years across much of Europe, California, and Canada.

Figures 7 and 8 of Chapter 11 (What is the difference between weather and climate?) show significant increases in global mean surface temperatures for the past 130 years or so. I don't think anyone can dispute the veracity of these increases in global mean temperatures.

What of the three other climate-related variables, drought, cyclones, and the ENSO Index?

Droughts (defined as significant and prolonged reductions in rainfall at a location/region compared to the long-term average) have been occurring for as long as continents have existed (that is, well before the evolution of humans). Droughts can be identified in tree rings of very old trees because droughts reduce annual tree growth and tree rings become narrower. The pertinent question here is: has the frequency of droughts increased in the past century (the period for which we can say global mean temperatures have increased)? The Palmer Drought Severity Index (PDSI) is an index of the severity of a drought in a specified region (see Textbox below). It is a simple model and is used extensively across the world for research and for practical applications in agricultural industries. Figure 11 is an example of the application of the PDSI across the USA.

Is there any evidence that the severity of drought, across the USA, for example, has increased over the past century? Apparently not. Analyses undertaken by the US Environmental Protection Agency demonstrate no change in the PDSI across the USA for the past 130 years (Fig. 12). In contrast, the IPCC AR6 report (published in 2022) states "Evidence of observed changes in extremes such as heatwaves, heavy precipitation, **droughts**, and tropical cyclones, and in particular, their attribution to human influence, has further strengthened since AR5 (the IPCC report published in 2014). Human influence has likely increased the chance of compound extreme events since the 1950s, including increases in the frequency of concurrent heatwaves and **droughts** (high confidence)." (from p5, section A.2.1 of the IPCC AR6 Summary Report).

Researchers in Italy recently analysed drought severity and frequency across southern, northern, and eastern Europe using the Standardized Precipitation Index (SPI), an alternative measure of drought to the PDSI (Spinoni et al., 2017, Global and Planetary Change) and concluded that drought severity and frequency **declined** across the period 1950 – 2014 for northern and eastern Europe, but there was a moderate increase for southern Europe. Increases in drought severity and frequency appear not to be universal, or even widespread.

Is there any evidence that the severity of tropical cyclones has increased over the past 70 years (when global mean temperature increase has been the largest)? A good integrated measure of the energy (hence severity), duration, and number of cyclones/hurricanes in an annual season is the accumulated cyclone energy (ACE) of the season. Figure 13 shows the results of the US Environmental Protection Agency analyses of ACE, and demonstrates that "year" explains almost none of the variation in ACE across the past 70 years and the regression is not statistically significant. Using the same data source, a plot of the number of cyclones in the north Atlantic between 1880 and 2018 yields a non-significant decline over that period, with an R^2 of 0.02 and a p-value of 0.77 (data not shown here). Yet the IPCC states "Other projected regional changes include intensification **of tropical cyclones and/ or extratropical storms (medium confidence)** and increases in aridity and fire weather (medium to high confidence)" (from p13, section B.1.4 of the IPCC AR6 Summary Report, 2022).

An explanation of the Palmer Drought Severity Index (PDSI)

The Palmer Drought Severity Index (PDSI) uses temperature and precipitation data to estimate relative dryness. It is a standardised index ranging from -10 (very dry) to +10 (very wet). The PDSI is reasonably successful at quantifying long-term drought. Because temperature data and a water balance model are used, it can capture the effects of global warming on droughts that occur via changes in atmospheric dryness. The figure below is an example of the use of the PDSI by NOAA to provide information to agriculturalists across the USA. A value of -4 reflects an extreme drought and a value of 4 reflects extremely high moisture availability.

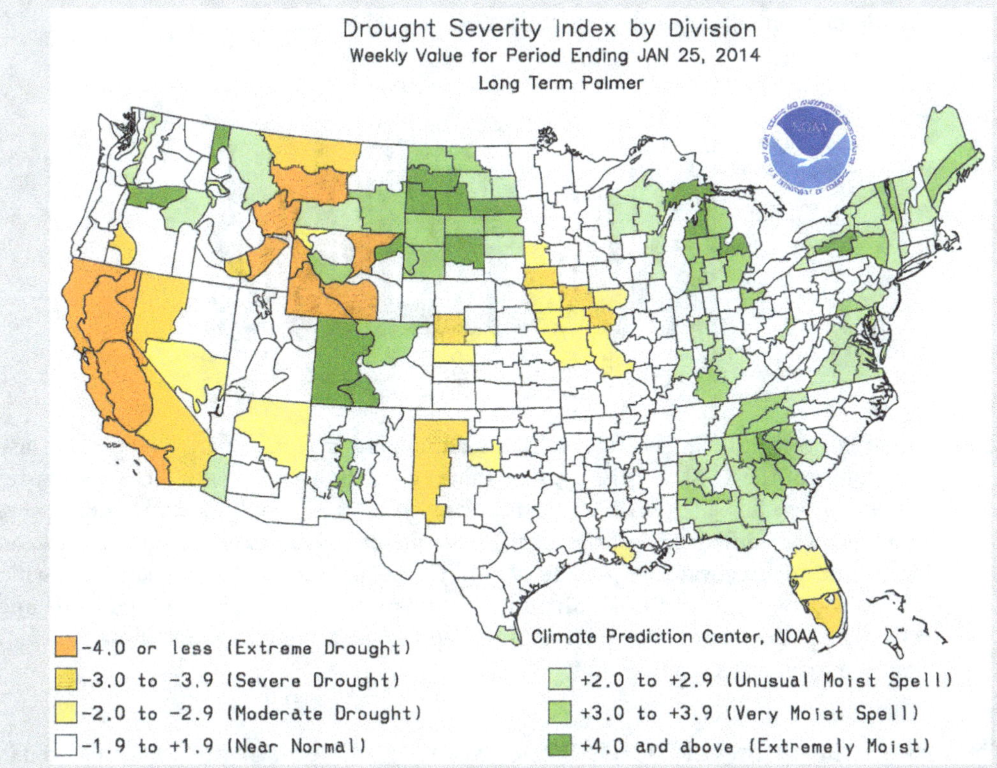

Figure 11: Long-term PDSI over the continental US through January 2014. From http://www.cpc.ncep.noaa.gov/products/analysis_monitoring/regional_monitoring. NOAA figures are available free of copyright.

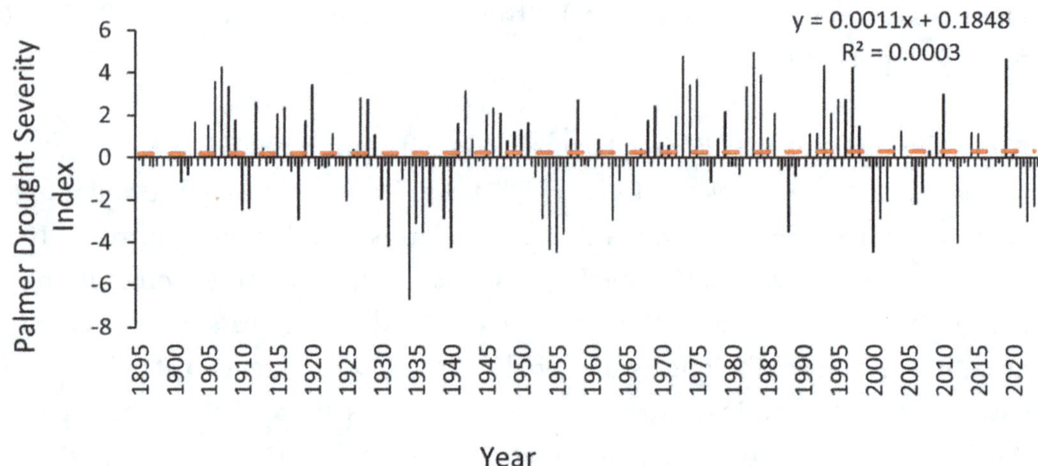

Figure 12: The Palmer Drought Severity Index for the USA shows no significant trend over the past 130 years. A regression through the annual data (red dashed line) yielded an R^2 of 0.0003 – a highly non-significant regression. Redrawn from: https://www.epa.gov/climate-indicators/climate-change-indicators-drought

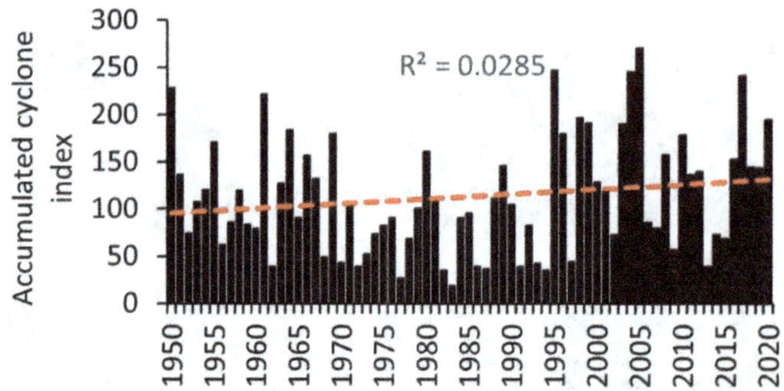

Figure 13: There is no significant change in the accumulated cyclone energy (expressed as a percentage of the 1981 – 2010 median value) over the past 70 years. The red dashed regression line has an R^2 of 0.0285; therefore "year" has very little explanatory power (it accounts for only 2.85% of the trend) in explaining changes in accumulated cyclone index. Put another way, the p-value of the regression is much larger than 0.05, which makes the regression non-significant. Redrawn from the EPA's Climate Change Indicators in the United States: https://www.epa.gov/climate-indicators/climate-change-indicators-tropical-cyclone-activity

This brings me to the ENSO (El Niño–Southern Oscillation). The ENSO is a natural weather/climate cycle that has existed for at least 13,000 years, but probably much longer (perhaps 45,000 years). ENSO has three phases: neutral, La Niña, and El Niño (Fig. 15). During the neutral phase the Humboldt current brings cold water from the Southern Ocean northwards up the western coast of South America. Equatorial winds drive this cold(ish) water west within the tropical region of the Pacific Ocean. As it moves west it is warmed by incoming solar radiation and sea-surface temperatures increase by 8 – 10 °C as it approaches the east coast of Australia. Perhaps somewhat surprisingly, the ocean is up to 60 cm higher in the western Pacific than the eastern Pacific after a couple of months of this wind-induced movement of water towards the western Pacific.

Evaporation from the warm surface water arriving in the western Pacific (i.e., on the east coast of Australia and Indonesia) increases cloud cover and rainfall in this region. As the air in the western Pacific rises (because of the warmth of the ocean surface warming the air), high altitude winds carry the now much drier air from the west to the east (back towards the west coast of South America). This circulation of air, from east to west at low altitude and from west to east at high altitudes, is known as the Walker circulation and is driven by gradients in atmospheric pressure between western and eastern regions of the Pacific.

The Southern Oscillation Index (SOI) is a measure of the difference in atmospheric pressure measured at Darwin (northern Australia) and Tahiti (an island midway between the east coast of Australia and the west coast of South America). When the index is smaller than -8 for two or more consecutive months, El Niño is likely, and when the SOI exceeds 8 for two or more consecutive months, La Niña is likely.

During El Niño years, the central and equatorial regions of the Pacific are warmer than usual because the westerly trade winds are weaker than usual and sea surface temperatures off the west coast of South America are warmer than during neutral years and sea surface temperatures are cooler close to Australia and Indonesia. The Walker circulation is reduced or absent during El Niño years. During La Niña years, the Walker circulation is enhanced and rainfall and temperature increase for eastern Australia but the opposite occurs along the west coast of South America.

The ENSO should be sensitive to climate change because of the potential for ocean currents (for example, the Humboldt current) to change in response to climate change, and the observed increase in sea surface temperatures (Fig. 14) that have occurred in the past century. Oscillations in the ENSO are associated with floods, droughts, and fires in Australia and South America.

Is there any evidence for a consistent change in the SOI over the past 130 years? Figure 15 suggests not.

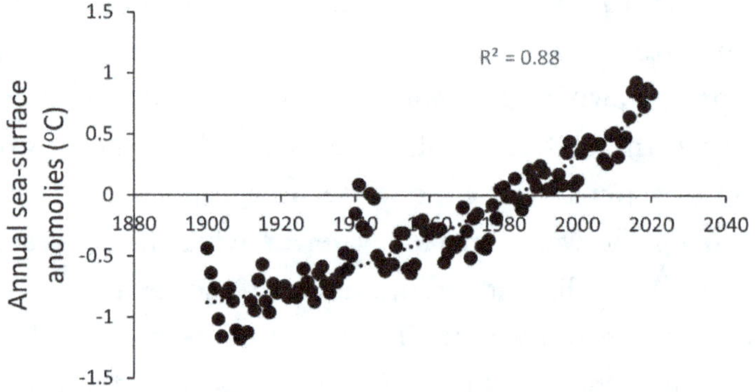

Figure 14: Average annual sea surface anomalies demonstrate a significant increase of about 1.6 °C since 1900. Redrawn from: https://www.epa.gov/climate-indicators/climate-change-indicators-sea-surface-temperature

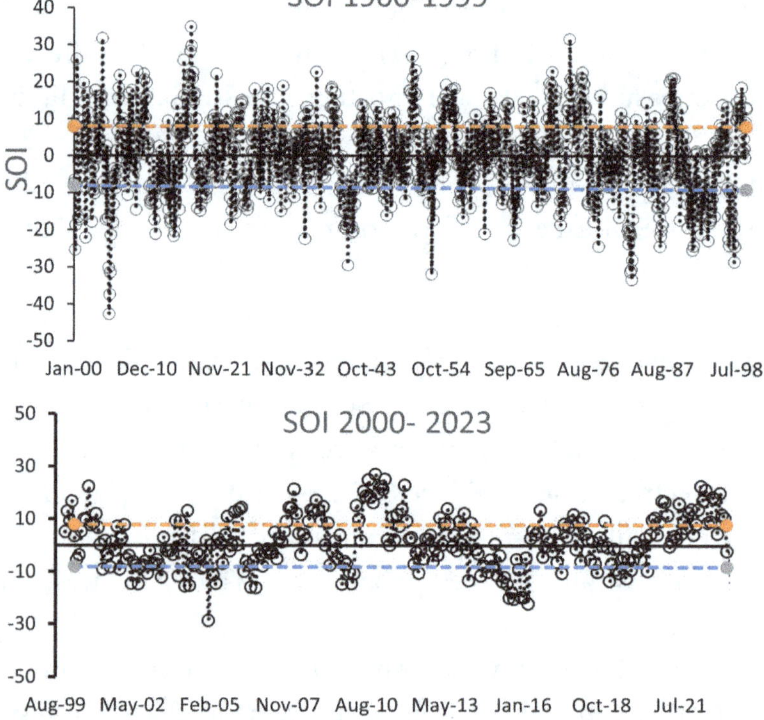

Figure 15: Upper panel: monthly SOI, 1900 - 1999. Jan-00 = Jan 1900, Jul-98 = Jul 1998. Lower panel: monthly SOI for 2000 – 2022. May-02 = May 2002, Jul-21 = Jul 2021.
Positive SOI >8 (red dashed line) – cool wet conditions for E. Australia = La Niña. Negative SOI <-8 (blue dashed line) = hot dry conditions for E. Australia = El Niño

Wildfires are associated with hot and dry conditions. Wildfires have been extensive and ferocious in California, Canada, and southern Europe across the past 3 – 6 years. Much has been made in the media of the increase in extent, ferocity, and frequency of such wildfires. Documenting these metrics is difficult, especially prior to the era of satellite remote sensing. However, several attempts at quantifying these have been made. A couple of examples suffice to illustrate the key point I wish to make.....

Figure 1. Annual Wildfires and Acres Burned, 1993-2022

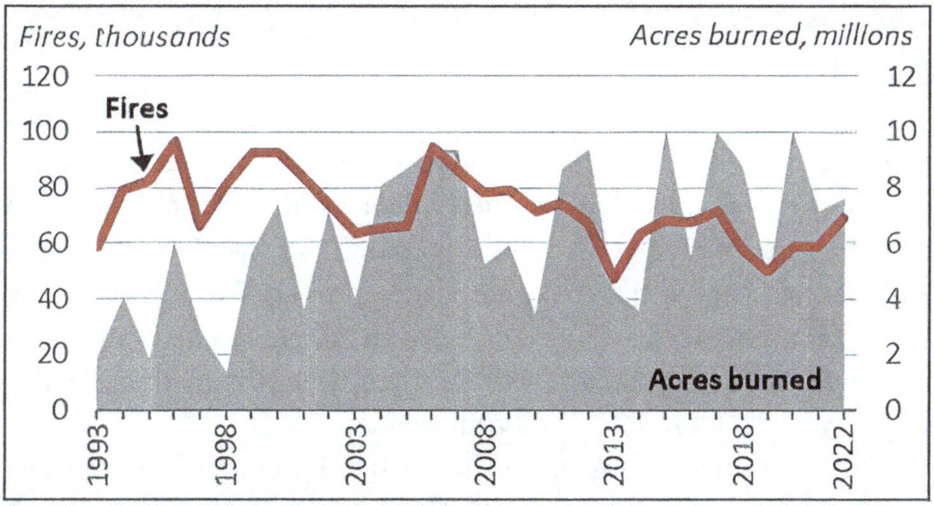

Source: NICC Wildland Fire Summary and Statistics annual reports.
Note: Data reflect wildland fires and acres burned nationwide, including wildland fires on federal and nonfederal lands.

Figure 16: The number of fires in continental USA between 1993 and 2022 have not increased, although the number of acres burned has increased significantly in that period.

Figure 16 (reproduced from a US Congressional Research Service report) demonstrates that the numbers of fires across continental USA has not increased in the past 39 years, although the number of acres burned has increased in that period. It has been suggested that the increase in the areal extent of forest reserves (a doubling since 1953) may have contributed to this increase in the total area of land burnt because forest reserves tend not to be managed as effectively as commercial/private forests to reduce fuel load. Are there any other data sources we can look to confirm or refute these data for the US?

A review of **global** trends in wildfires was published in 2016, in the prestigious journal Philosophical Transactions of the Royal Society B, (https://royalsociety.org/blog/2020/10/global -trends-wildfire/). A single quote may be sufficient here:

"….*global area burned appears to have overall declined over past decades*, and there is increasing evidence that *there is less fire in the global landscape today than centuries ago.* Regarding fire severity, limited data are available. For the western USA, they indicate little change overall, and that the area burned at high severity has overall declined compared to pre-European settlement." (Doerr and Santin, 2016, Philosophical Transactions of the Royal Society B).

Similar absences of a consistent increase in numbers of fires and/or the area burnt have been published for the European Mediterranean region (Fig. 17) and in the American west (Marlon et al., 2012, Proceedings of the National Academy of Sciences), in which rates of charcoal were used to produce fire histories going back about 3000 years. These researchers concluded that there has been a small decline in burning over the past 3000 years, with the lowest levels attained during the Little Ice Age (*ca.* 1400 – 1700 AD) and the 20th century.

It would be remiss of me to not mention that there are several studies that suggest an increase in fire frequency and/or intensity and/or area burned has occurred in the recent past. It seems very likely to me that continued increases in land surface temperature coupled to increased aridity *in some regions* will tend, in the short-term, to increase the area of land burned. Indeed, the IPCC AR6 report states "There is medium confidence that weather conditions that promote wildfires have become more probable in southern Europe, northern Eurasia, the USA, and Australia over the last century. (Box TS.10 P108)". Whether increased frequencies of burning could be maintained for long is debatable because the lack of fuel, arising from the lack of water to support plant growth, won't support extreme fires indefinitely. The two key conclusions from this discussion are, first, media hype of recent fire storms and climate change are not supported by data. Second, we don't have enough data to conclusively demonstrate an upward trend in fire frequency yet.

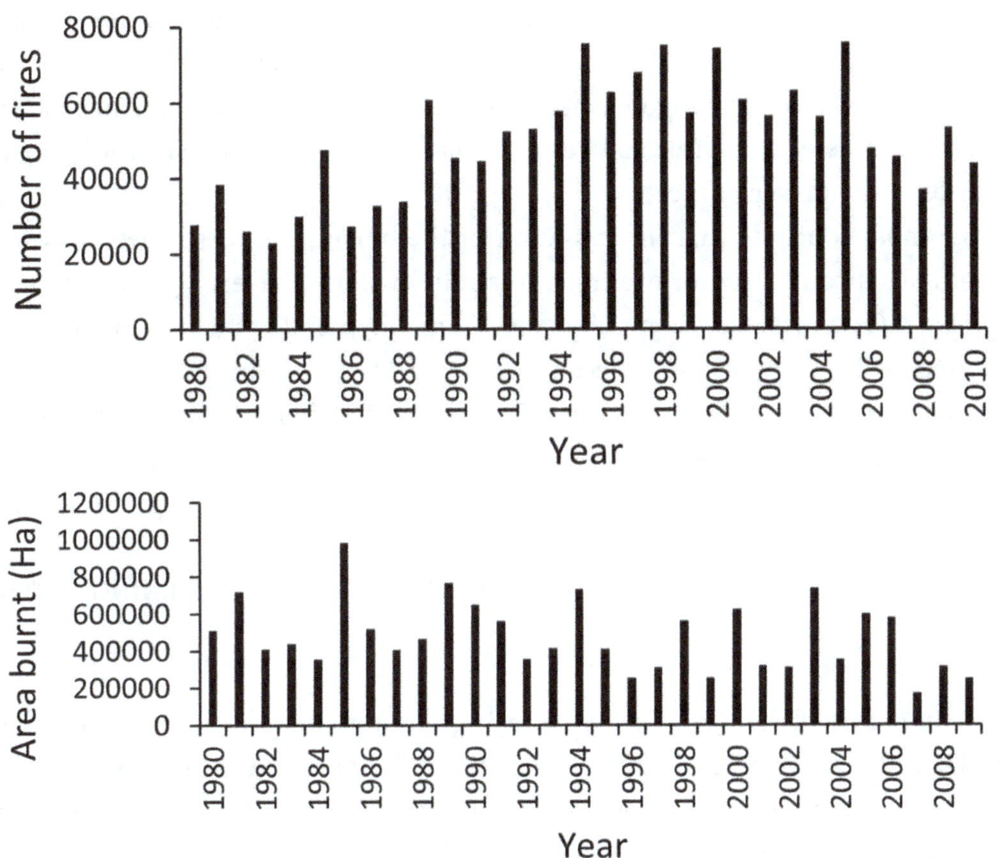

Figure 17: Analyses of fire data in European Mediterranean landscapes did not find evidence of a consistent increase in either the number of fires (upper) or the area burnt (lower), between 1980 and 2010. Redrawn from: San-Miguel-Ayanz et al. (2013) Analysis of large fires in European Mediterranean landscapes: lessons learned and perspectives. For. Ecol. Manage. 294, 11– 22. (doi:10.1016/j.foreco.2012.10.050).

So, **finally, to hypothesis 3** of Figure 5, which states that stopping the burning of fossil fuels is the optimal solution to the problem of climate change and is rational, economically justified, and feasible from both an engineering and strategic geopolitical viewpoint.

If the burning of fossil fuels is leading to climate change, and the impacts of climate change are disastrous for the world (because of increased droughts, cyclones, species extinctions, wildfires, ENSO events, catastrophic increases in sea-levels; but see Figs. 12, 13, 15, and 16) then the global cessation of that burning would be a rational response. Australia and several other countries have committed themselves to a zero-emission economy by 2050 and a complete phasing out of, for example, generating electricity using coal-fired power stations. However, statistics presented on pages 246-247 concerning Germany, South Africa, China, India, and the USA, suggest that this is not a rational response by Australia and most other Western countries at the present time.

One must conclude that the cessation of the burning of fossil fuels by several medium-sized countries (for example, Australia) will not have a significant impact on global CO_2 emissions. Remember too, from Chapter 11, that when CO_2 emissions declined by 5 % in 2022 because of COVID-19, the inexorable increase in atmospheric CO_2 concentrations did not miss a beat. Therefore, atmospheric CO_2 concentrations will probably not be significantly reduced by the headlong rush to decarbonise Western economies and thus this is an irrational strategy to pursue.

COST-BENEFIT ANALYSES OF THE STRATEGY OF CESSATION OF BURNING FOSSIL FUELS

A well understood framework for choosing among competing strategies in the allocation of resources is to apply a cost-benefit analysis. A cost-benefit analysis (CBA) uses a systematic approach, often, but not always, an economic approach, to estimating the strengths and weaknesses of alternative strategies. It can identify which strategy provides the best approach to achieving maximal benefits for minimal cost. It can be used in business and public policy fora. It is too infrequently used in the climate change policy space. Bjorn Lomberg of The Copenhagen Consensus Centre at The Hoover Institute, Stanford University, is rigorously attempting to improve the application of CBA within the climate change policy space. He collaborates with multiple Nobel Laureate economists to apply CBA to the world's biggest social and ecological problems, including climate change policy analyses. In 2020 he published a review of the socio-economic impacts of alternate strategies for coping with climate change. Some of his conclusions are:

1. Predictions of environmentally catastrophic results arising from climate change are almost entirely unwarranted. Unfortunately, current climate mitigation policies are based on such unwarranted predictions and fail simple cost-benefit analyses.

2. A good measure of human wellbeing is GDP/capita, where GDP is gross domestic product. While not perfect as an index, it does correlate strongly with life expectancy, education attainment and the UN's Human Development Index. Higher GDP/capita is correlated with reductions in local pollution, improved access to sanitation, electricity and other infrastructure, improved nutrition, and escape from poverty.

3. The best available estimates of the total impact of elevated global mean surface temperatures for the current 1 – 1.5 °C rise (since 1900) on the decline in global GDP (expressed as a percentage of GDP) ranges between zero and a small (< 0.15 %) decline. Even a 4 °C increase in temperature results in only a 2.9 % decline in GDP. By building in an extra 25 % in losses to account for losses that may occur, but which have yet to be quantified, this results in a decline of GDP of 3.64 %. Is this a catastrophic loss of global GDP? No, it isn't, because we need to remember that GDP will continue to increase throughout the 21st century. The real question is, how much less will GDP have grown by 2100 if climate impacts on GDP are 3.64 % by 2100?

4. In the sustainable futures scenario model of the IPCC, by 2100, the per person GDP will have grown 6 times larger if there was no negative impact of climate change during the 21st century, but it will have grown by 5.9 times if the climate impact is 2.5 % of GDP. Alternatively, in the "no change, continue burning fossil fuels" scenario, GDP would increase by 10.4 times its 2020 value with no impacts on climate change, or GDP per person will grow by 9.8 times its 2020 value if the impact of climate change is a 5.7 % decline in per person GDP. **The key take-home message here is that the impact of climate change on per person GDP is very small.**

5. The dollar cost of implementing the Paris Agreement greatly exceeds the dollar benefits of implementing the policy responses within the Paris Agreement. For every dollar spent on implementing the Paris Agreement, 11 cents of benefit accrue.

6. The costs of doing nothing with respect to mitigating negative impacts of climate change on per capita GDP are discussed extensively and repeatedly in the media. What is not discussed is that implementing climate policies also incur a massive cost and this cost often exceeds the benefits of implementing a specific policy.

7. It is possible to estimate the dollar benefit of spending on strategies to meet the UN's Sustainable Development Goals (for example, eradicating malaria, extending mobile broadband cover in developing counties, supplying modern cooking fuels to the developing world, extending the supply of reliable electricity to all the developing world, increasing access to education for girls). Achieving all these goals generate a benefit that exceeds the cost (Lomberg, 2020, Technological Forecasting and Social Change). The ratio of benefits to costs ranges from $2011 (that is, for every $1 spent on a strategy, $2011 benefit accrues) for removing all trade barriers, to $36 for halving the incidence of malaria, or $5 for supplying electricity to all the developing world. However, keeping global mean surface temperature rise to less than 2 °C **results in a cost that exceeds the benefit** and therefore chasing that goal is an irrational choice.

Let me emphasise that the ratio of dollar values of costs to benefits are contested numbers and the methodology (what to include, what to exclude, for example) is also contested. I am not able to say that these numbers are correct, approximately correct, or very wrong. My point of including this discussion is to highlight that cost/benefit analyses rarely appear in media reports and this is a serious omission. Decisions of such magnitude (rewriting the economic basis to the global economy; changing the power relationships between the democratic West and autocratic parts of the east) require serious consideration of the optimisation of allocation of limited resources. It is rather disappointing to note that a recent attempt by Lomberg to establish a Climate Change Centre at an Australian University to undertake such cost/benefit analyses was thwarted by a number of branches of student unions and the Australian academic union.

A MORAL AND A GEOSTRATEGIC PERSPECTIVE

There is a strong moral argument put forward by many developing countries (including China (how is China a developing country?), India, Indonesia, and most African and Pacific island nations) that Europe, Australia and the USA currently enjoy the lifestyle and quality of life that they do because they were able to chop down forests and burn vast amounts of fossil fuels for two centuries, unhindered by thoughts of climate change. Therefore, it is not acceptable (so the argument goes) for Europe and the USA to prohibit the cutting of forests and the burning of fossil fuels in developing countries, which (so the argument goes) should be allowed to achieve a similar lifestyle and *quality of life* as the developed world and enjoy the *longevity of lifespan and wealth enjoyed by the developed world*. Put another way, and as discussed by Lomberg in his 2020 review, global inequality (in wealth distribution, educational attainment, health, lifespan) can only be reduced if the developing world has the same access to cheap energy as was enjoyed by Europe and the USA for 250 years after the Industrial Revolution.

From a geostrategic perspective, it has been argued that the decarbonisation of the developed world will seriously, and potentially catastrophically, place the security of the developed world in such a weak position that countries with authoritarian and despotic political systems, which will not decarbonise, will be able to change the organisation of global trade and global power relationships away from the "rules-based" order that has guaranteed an unprecedented 7 decades of continental- and inter-continental-scale peace, and an almost continuous growth in global GDP.

Can we confidently and safely run the economies of developed countries when the means of production and distribution of key metals, including lithium, cobalt, and rare earth metals (for example, neodymium, samarium, terbium, and dysprosium), is centralised in authoritarian and despotic countries? Such metals are required in increasingly large amounts for rechargeable batteries, the magnets in electric vehicles, and the vast storage batteries required to smooth-out the supply of electricity at a regional-scale when renewables are a large fraction of the electricity supply mix.

In conclusion, this chapter has, I hope, demonstrated the following:

1. Theories, hypotheses, and Laws are not the same thing, but share some similarities.
2. A strong correlation between two variables does not, of itself, prove a causal relationship between those variables.
3. The Scientific Method is a process, or a set of linked processes, but there is not a single universally applicable Scientific Method.

4. Science can, occasionally, in limited and well-defined cases, "be settled" but the Science of Climate Change is certainly not settled and won't be in the near-to-mid-term time-scales, and probably never.

5. It is undoubtedly true that sea-surface temperatures and land-surface temperatures have increased significantly in the past 100+ years. Simultaneously, atmospheric CO_2 levels have increased significantly through burning of fossil fuels and deforestation.

6. There is no evidence of consistent upward trends in drought, wildfire, cyclonic intensification, or increased frequency or intensity to the ENSO (El Niño, La Niña) over the past 100 years or so.

7. Cost-benefit analyses of the UN's Sustainable Development Goals emphasise that while most goals yield a benefit exceeding the cost (more dollars accrued than spent), the cost of limiting the increase of global mean temperatures to less than 2 °C results in a benefit smaller than the cost. There are more rational ways to spend limited funds to improve global welfare, including investment in mitigation strategies to cope with increased surface temperatures. The aim of strategies to tackle climate change should be to maximise the welfare of the largest number of people, globally.

THE END